Black Holes and the Structure of the Universe

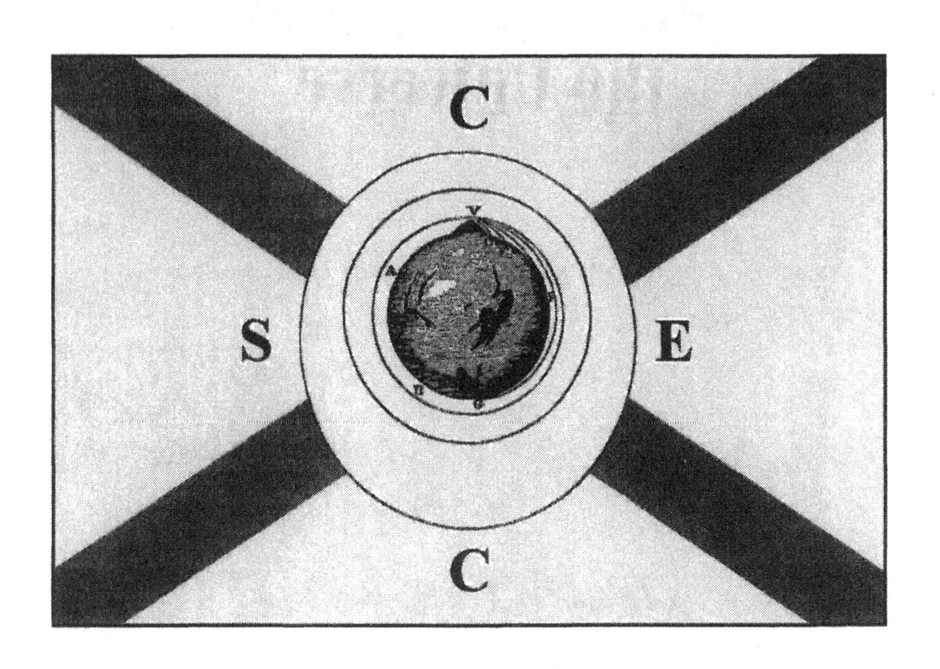

Published by World Scientific Publishing Co. in association
with the Centro de Estudios Científicos (CECS), Valdivia, Chile

Black Holes and the Structure of the Universe

Editors

Claudio Teitelboim
Jorge Zanelli
Centro de Estudios Científicos (CECS)
Valdivia, Chile

Published by

World Scientific Publishing Co. Pte. Ltd.

P O Box 128, Farrer Road, Singapore 912805

USA office: Suite 1B, 1060 Main Street, River Edge, NJ 07661

UK office: 57 Shelton Street, Covent Garden, London WC2H 9HE

British Library Cataloguing-in-Publication Data
A catalogue record for this book is available from the British Library.

BLACK HOLES AND THE STRUCTURE OF THE UNIVERSE

ISBN 981-02-4269-7

Printed in Singapore.

CONTENTS

PREFACE

The extreme geography of Chile and of Antarctica were the settings for sessions of the meeting on "Black Holes and the Structure of the Universe", from which this volume stems.

The meeting, held in August 1997, was made possible by a joint effort in which the government, the private sector and the Chilean Air Force, among others, took part. This common will for the cause of science and the understanding of the Universe meant much for the society at large in a country still confronting divisions from the past. To all those who made it possible, especially the participants and the speakers who came from all over the world, we extend our profound gratitude.

C. Teitelboim and J. Zanelli

HIGHER DIMENSIONAL CHERN-SIMONS THEORIES AND TOPOLOGICAL BLACK HOLES

MÁXIMO BAÑADOS

Centro de Estudios Científicos de Santiago, Casilla 16443, Santiago, Chile, and Departamento de Física, Universidad de Santiago de Chile, Casilla 307, Santiago 2, Chile

It has been recently pointed out that black holes of constant curvature with a "chronological singularity" can be constructed in any spacetime dimension. These black holes share many common properties with the 2+1 black hole. In this contribution we give a brief summary of these new black holes and consider them as solutions of a Chern-Simons gravity theory. We also provide a brief introduction to some aspects of higher dimensional Chern-Simons theories.

1 The topological black hole

A topological black hole in n dimensions can be constructed by making identifications along a particular Killing vector on n dimensional anti-de Sitter space, just as the 2+1 black hole is constructed from 3 dimensional anti-de Sitter space. This procedure can be summarized as follows. Consider the n dimensional anti-de Sitter space

$$-x_0^2 + x_1^2 + \cdots + x_{n-2}^2 + x_{n-1}^2 - x_n^2 = -l^2, \tag{1}$$

and consider the boost $\xi = (r_+/l)(x_{n-1}\partial_n + x_n\partial_{n-1})$ with norm $\xi^2 = (r_+^2/l^2)(-x_{n-1}^2 + x_n^2)$. For $\xi^2 = r_+^2$, one has the null surface,

$$x_0^2 = x_1^2 + \cdots + x_{n-2}^2, \tag{2}$$

while for $\xi^2 = 0$ one has the hyperboloid,

$$x_0^2 = x_1^2 + \cdots + x_{n-2}^2 + l^2. \tag{3}$$

Let us now identify points along the orbit of ξ. The region behind the hyperboloid ($\xi^2 < 0$) has to be removed from the physical spacetime because it contains closed timelike curves. The hyperboloid is thus a singularity because timelike geodesics end there. On the other hand, the null surface (2) acts as a horizon because any physical observer that crosses it cannot go back to the region which is connected to infinity. Indeed, the surface (2) coincides with the boundary of the causal past of light like infinity. In this sense, the surface (1) with identified points represents a black hole. The existence of this n dimensional anti-de Sitter black hole was first pointed out in [1], in four dimensions.

Let us now introduce local coordinates on anti-de Sitter space (in the region $\xi^2 > 0$) adapted to the Killing vector ξ. We introduce the n dimensionless local coordinates (y_α, ϕ) by,

$$
\begin{aligned}
x_\alpha &= \frac{2ly_\alpha}{1-y^2}, \qquad\qquad \alpha = 0, ..., n-2 \\
x_{n-1} &= \frac{lr}{r_+}\sinh\left(\frac{r_+\phi}{l}\right), \\
x_n &= \frac{lr}{r_+}\cosh\left(\frac{r_+\phi}{l}\right),
\end{aligned}
$$

with $r = r_+(1+y^2)/(1-y^2)$ and $y^2 = \eta_{\alpha\beta}\, y^\alpha y^\beta$ $[\eta_{\alpha\beta} = \mathrm{diag}(-1, 1, ..., 1)]$. The coordinate ranges are $-\infty < \phi < \infty$ and $-\infty < y^\alpha < \infty$ with the restriction $-1 < y^2 < 1$. The induced metric has the Kruskal form,

$$
ds^2 = \frac{l^2(r+r_+)^2}{r_+^2} dy^\alpha dy^\beta \eta_{\alpha\beta} + r^2 d\phi^2, \tag{4}
$$

and the Killing vector reads $\xi = \partial_\phi$ with $\xi^2 = r^2$. In these coordinates, the quotient space is simply obtained by identifying $\phi \sim \phi + 2\pi n$, and the resulting topology is $\Re^{n-1} \times S_1$. The metric (4) represents the n-dimensional topological black hole black hole written in Kruskal coordinates. Note that the above metric is a natural generalization of the 2+1 black hole. Indeed, setting $n = 3$ in (4) gives the non-rotating 2+1 black hole metric written in Kruskal coordinates [2].

The fact that the metric (4) is a natural generalization of the 2+1 black hole is of special relevance. Recently the 2+1 black hole has been shown to provide interesting applications in string theory [3].

An important characteristic of the topological black hole for $n > 3$ is the non-existence of a globally well defined timelike Killing vector. In other words, the black hole manifold is not static. This is easily seen by studying the behavior of the anti-de Sitter Killing vectors after the identifications are done [4]. It is possible, however, to choose local static coordinates that resembles the Schwarszchild coordinates [5].

This problem does not appear when studying the "Euclidean black hole". If we consider Euclidean anti-de Sitter space and then make identifications one can produce a non-trivial manifold that can be called the Euclidean topological black hole. Although in the Euclidean sector the notion of timelike does not make sense, one can identify a "timelike" Killing vector ∂_t which is globally defined.

By construction, the topological black hole has constant negative curvature and therefore it solves Einstein's equations with a negative cosmological constant. However, for $n > 3$, the ADM mass is infinity (see [6] for the explicit calculations). Here, we mean the ADM mass for the Euclidean black hole. It was pointed out in [5] that finite conserved charges for the topological black hole can be defined in the context of a Chern-Simons gravitational theory. In the next section we shall give a brief introduction to higher dimensional Chern-Simons theories and then consider the particular case of Chern-Simons gravity.

2 Chern-Simons theory

2.1 The Lagrangian equations of motion

A Chern-Simons theory can be regarded as a Yang-Mills theory with an exotic action. The main ingredients of a Chern-Simons theory in n dimensions are; a N dimensional Lie algebra with generators T_a satisfying $[T_a, T_b] = f^c{}_{ab}T_c$ ($a = 1, ..., N$), a Yang-Mills gauge field $A = A^a T_a$, and a $n-1$ totally symmetric invariant tensor $< T_{a_1}, ..., T_{a_{n-1}} > \equiv g_{a_1 a_2 ... a_{n-1}}$.

The Chern-Simons equations of motion can then be written as

$$g_{a_1 ... a_{n-1}} F^{a_2} \wedge F^{a_2} \wedge \cdots F^{a_{n-1}} = 0 \tag{5}$$

where $F^a = dA^a + (1/2) f^a_{bc} A^b \wedge A^c$. These equations of motion are derived from an action principle with a Lagrangian L that satisfies $dL = g_{a_1 ... a_{n-1}} F^{a_1} \wedge \cdots \wedge F^{a_{n-1}}$.

In the three dimensional case, these equations reduce to $g_{ab} F^b = 0$. If g_{ab} is non-degenerate, then they simply imply $F^a = 0$ and the theory reduces to the problem of studying the class of flat connections modulo gauge transformations. The space of solutions of the equations of motion is thus completely classified given the topology of the manifold. An immediate consequence of this is that there are no local degrees of freedom in three dimensional Chern-Simons theory.

In higher dimensions, however, the set of equations (5) are far more complicated; they possess local degrees of freedom [7] and the space of solutions cannot be associated uniquely to the spacetime topology. Furthermore, the phase space is stratified in regions with different number of degrees of freedom. The maximum number of local physical degrees of freedom is equal to $mN - N - m$, with $n = 2m + 1$. This formula is valid only for $n > 3$ and $N > 1$. [a]

---

[a] For $N = 1$ the Chern-Simons theory is surprinsingly more complicated because the separation of first nd second class constraints cannot be achieved in a generally covariant form. The $N = 1$ theory in any number of dimensions does not have any local degrees of freedom.

Despite the complicated nature of the equations (5), a full Hamiltonian decomposition can be performed [7], and in this form the equations take a simple form. Here we shall consider the equations of motion in a space+time form, in five dimensions. The general situation as well as the Hamiltonian structure is analysed in [8].

2.2 Space + time decomposition

Suppose that locally we decompose the gauge field as

$$A^a_\mu dx^\mu = A_0 dt + A_i dx^i, \tag{6}$$

then the above equations, for $n = 5$, can be split in the $4 + 1$ form

$$\epsilon^{ijkl} g_{abc} F^b{}_{ij} F^c{}_{kl} = 0, \tag{7}$$

$$\Omega^{ij}_{ab} F^b_{j0} = 0, \tag{8}$$

with

$$\Omega^{ij}_{ab} = \epsilon^{ijkl} g_{abc} F^c{}_{kl}. \tag{9}$$

Note that contrary to the Yang-Mills equations of motion, these equations involve only the curvature tensor, and can be regarded as an algebraic set of equations for $F^a_{\mu\nu}$. Thus, the integration of the higher dimensional Chern-Simons theory equations is equivalent to an algebraic problem plus solving the Bianchi identities $DF^a = 0$.

Equations (7) do not have any time derivatives and therefore they are constraints over the initial data. To our knowledge, the space of solutions of these equations is not known. Equations (8), on the other hand, do contain time derivatives. However, it is not obvious that there are no further constraints among them.

The nature of the above equations is completely governed by the algebraic properties of the $4N \times 4N$ matrix Ω^{ab}_{ij}, which depends on the invariant tensor g_{abc} and F^a_{ij}. Indeed, Eq. (8) depends explicitly on Ω while, using some simple combinatorial properties, Eq. (7) can also be written in terms of Ω as

$$\Omega^{ij}_{ab} F^b_{jk} = 0. \tag{10}$$

In this form, the constraint is equivalent to the statement that F^a_{ij} must be a zero eigenvalue of Ω^{ij}_{ab}. It has been shown in [7,8] with many examples that generically there exists solutions to the constraint equations (7) for which the only zero eigenvectors of Ω are precisely F^a_{ij}. That is, if V^a_i satisfies $\Omega^{ij}_{ab} V^b_i = 0$, then there exists a vector field N^i such that $V^a_i = F^a_{ij} N^j$. The matrix Ω thus has four, and only four, zero eigenvalues.

The space of solutions of the constraint satisfying this property carry the maximum number of degrees of freedom and we shall consider here only this sector of the theory. Note that, in particular, we exclude the flat solutions $F^a_{ij} = 0$.

We now turn to the dynamical equations. Equations (8) imply that F^a_{i0} is a zero eigenvector of Ω^{ij}_{ab}. The above discussion thus leads to the existence of a "shift" vector N^i such that

$$F^a_{i0} = F^a_{ij} N^j. \tag{11}$$

Noting that $F^a_{i0} = \dot{A}^a_i - D_i A^a_0$ this equation is equivalent to the statement that the time evolution is generated by a gauge transformation with parameter A^a_0 plus a spatial diffeomorphism with parameter N^i. The appearence of the spatial diffeomorphisms in the dynamical evolution reflects the fact that the gauge field is not flat and diffeomorphisms cannot be absorbed in the group of gauge transformations.

In the "time gauge" $A^a_0 = 0$ and $N^i = 0$, the dynamical equations simply imply $\dot{A}^a_i = 0$ and therefore one is left only with the constraint equation (7). We shall see below that for the particular group $SO(4,2)$, the above equations of motion represent the generalization of the Einstein equations in five dimensions due to Lovelock. Thus, if the constraint (7) was integrable, that would imply the integrability of the Chern-Simons Lovelock theory of gravity.

2.3 The $G \times U(1)$ theory

We have seen in the last section that on the space of solutions for which Ω has the maximum rank, the equations of motion can be unambigously separated into constraints plus dynamical equations. However, we have not yet proved that the condition that Ω has only 4 null eigenvalues is not empty.

The maximum rank condition can be explicitely implemented in a remarkably simple form if we couple to the original Chern-Simons action an Abelian $U(1)$ field, that we shall call b [8]. As a matter of fact, we shall see below that an Abelian field with the correct coupling appears naturally in five dimensional Chern-Simons Supergravity.

If we add to the original action the term $b \wedge F^a \wedge F^b g_{ab}$ with g_{ab} the Killing form of the Lie algebra G, then the equations of motion are modified as

$$g_{abc} F^b \wedge F^c = H \wedge F^b g_{ab}, \qquad F^a \wedge F^b g_{ab} = 0, \tag{12}$$

where $H = db$ is the field strength of the Abelian field. These equations represent a Chern-Simons theory for the group $G \times U(1)$. Indeed, it is a simple exercise to prove that if we collect together the gauge field $A^A = (b, A^a)$, then there exists an invariant tensor g_{ABC} of $G\times U(1)$ such that the above equations can be written as $g_{ABC}F^B{}_\wedge F^C = 0$.

The usefulness of coupling the Abelian field b is that now the maximum rank condition can be achieved simply by imposing that the pull back of H to the spatial surface must be non-degenerate, that is $\det(H_{ij}) \neq 0$. To see this first note that the equations (12) are solved by $F^a = 0$ and H arbitrary. Second, the matrix Ω evaluated on this particular solution has the block form

$$\Omega^{ij}_{AB}\Big|_{F^a=0} = \left(\begin{array}{c|c} 0_{4\mathrm{x}4} & 0_{4\mathrm{x}4N} \\ \hline 0_{4N\mathrm{x}4} & g_{ab}\epsilon^{ijkl}H_{kl} \end{array}\right). \tag{13}$$

The 4x4 zero block provides exactly the zero eigenvalues associated to the null eigenvectors F^a_{ij} of Ω. On the other hand, imposing H_{ij} and g_{ab} to be non-degenerate, the lower $4N$x$4N$ block is non-degenerate and Ω has indeed only four zero eigenvalues.

We can now make perturbations with respect to this background. Since a non-zero eigenvalue cannot be set equal to zero by a small perturbation, the maximum rank condition is stable under small perturbations.

2.4 The WZW_4 algebra

Perhaps the most interesting application of higher dimensional $G \times U(1)$ Chern-Simons theories is its relation with the WZW_4 theory proposed in [9], and further developed in [10]. These theories are generalizations of the standard two dimensional WZW theories.

Let us first briefly review the relation between three dimensional Chern-Simons theory and the two dimensional WZW model in the form developed in [11]. Due to the non-existence of local degrees of freedom in 3D Chern-Simons theory, one can solve the constraint $F^a_{ij} = 0$ as $A_i = g^{-1}\partial_i g$, where g is a map from the manifold to the group. Replacing back this value of A_i into the Chern-Simons action one finds a chiral WZW action for the map g. The simplectic structure of the WZW model implies that the current $J(\lambda) = \int_{\partial\Sigma} Tr(\lambda A)$ satisfies the one dimensional Kac-Moody algebra [12]. A different way to arrive at the same result is by studying the issue of global charges [13] in the Chern-Simons action. Indeed, if the Chern-Simons theory is formulated on a manifold with a boundary, then one can show that under appropriate boundary conditions, there exists an infinite set of global charges equal to $J(\lambda)$ that satisfy the Kac-Moody algebra.

In the five dimensional $G\times U(1)$ Chern-Simons theory, the solution $F^a_{ij} = 0$ to the constraint is by far not the most general one, although it is a good background in the sense that it carries the maximum number of degrees of freedom. Due to the existence of local degrees of freedom, one cannot solve the constraints in a close and general form, and therefore one does not find a simple model at the boundary. Still, one can study the issue of global charges and impose as a boundary condition that A^a must be flat. This has been done in detail in [8]. One finds an infinite tower of global charges given by $J(\lambda) = \int_{\partial\Sigma} H_\wedge Tr(\lambda A)$ and they satisfy the extension to three dimensions of the Kac-Moody algebra.

3 Five dimensional Chern-Simons Gravity

3.1 The action

It is well known that in dimensions greater than four the Hilbert action is no longer the most general action for the gravitational field. For $D > 4$, there exists a class of tensor densities that, as the Hilbert term, give rise to second order field equations for the metric, and a conserved energy momentum tensor[14]. These terms are proportional to the dimensional continuation of the Euler characteristic of all dimensions $2p < D$ [15].

For odd dimensional spacetimes there exists a particular combination of those terms such that the resulting theory can be regarded as a Chern-Simons theory of the form described in the last section[16]. Black holes solutions for this theory were found in [18].

The simplest Chern-Simons theory of gravity exists in 2+1 dimensions with action

$$I_{2+1} = \int \epsilon_{abc} R^{ab}{}_\wedge e^c. \tag{14}$$

This action can be regarded as a Chern-Simons theory for the Poincaré group. Indeed, besides local Lorentz rotations, (14) is also invariant under Poincaré translations,

$$\delta e^a = D\lambda^a, \qquad \delta w^{ab} = 0 \tag{15}$$

as can be easily verified using the Bianchi identity $DR^{ab} = 0$. It is a simple exercise to prove that the 3+1 Hilbert counterpart of (14) is not invariant under this transformation.

The action (14) can be extended to any odd dimensional spacetime. For example, in five dimensions we consider

$$I_{4+1} = \int \epsilon_{abcde} R^{ab}{}_\wedge R^{cd}{}_\wedge e^e. \tag{16}$$

which is also invariant under (15). The key property that (14) and (16) share and makes them invariant under (15) is that they are linear in the veilbein e^a. This is also the property that makes (16) a Chern-Simons theory in five dimensions. Just as (14), the action (16) has a simple supersymmetric extension [19].

The actions (14) and (16) can be deformed to include a cosmological constant. For example, in five dimensions, the anti-de Sitter Chern-Simons theory of gravity is described by an action

$$I_{4+1}^{\Lambda} = \int \epsilon_{abcde}(R^{ab}{}_{\wedge}R^{cd}{}_{\wedge}e^e + \frac{2}{3l^2}R^{ab}{}_{\wedge}e^c{}_{\wedge}e^d{}_{\wedge}e^e + \frac{1}{5l^4}e^a{}_{\wedge}e^b{}_{\wedge}e^c{}_{\wedge}e^d{}_{\wedge}e^e) \quad (17)$$

where l is a parameter with dimensions of length that parametrizes the cosmological constant. Note that apart from an overall constant (Newton's constant) and l, there are no other free parameters in this action. Apart from the explicit local Lorentz invariance, the action (17) is also invariant under the deformed version of (15),

$$\delta e^a = D\lambda^a, \qquad \delta w^{ab} = \frac{1}{l^2}(e^a\lambda^b - e^b\lambda^a) \quad (18)$$

that reduces to (15) for $l^2 \to \infty$. As in the Poincaré case, this action can be made supersymmetric (see [16,17] and the contribution by J. Zanelli in this volumen) in a simple way.

The transformations (18) plus the Lorentz rotations form a representation of the orthogonal group $SO(4,2)$. Although the action is not explicitly invariant under this larger symmetry, the equations of motion following from (17) can be collected as

$$\epsilon_{ABCDEF}\tilde{R}^{AB}{}_{\wedge}\tilde{R}^{CD} = 0 \quad (19)$$

with $A = (a,6)$, $\tilde{R}^{ab} = R^{ab} + (1/l^2)e^a{}_{\wedge}e^b$ and $\tilde{R}^{a6} = (1/l)T^a$. In this form, the $SO(4,2)$ symmetry is explicit.

3.2 Black holes

An interesting property of the action (17) is the existence of two different black hole solutions for its equations of motion. One the one hand, there exists the topological black holes described in the first section with topology $\Re^4 \times S_1$, constant curvature (zero anti-de Sitter curvature $R^{AB} = 0$), and a chronological singularity. On the other hand, the line element

$$ds^2 = -N^2dt^2 + N^{-2}dr^2 + r^2d\Omega_3 \quad (20)$$

with

$$N^2 = 1 - \sqrt{M+1} + \frac{r^2}{l^2} \tag{21}$$

is also an exact solution of (17) [18]. The constant M is the ADM mass of the solution and one can see that an horizon exists only for $M > 0$. For $M = -1$ one has anti-de Sitter space. The scalar curvature of this metric is equal to

$$R = -\frac{20}{l^2} + \frac{6\sqrt{M+1}}{r^2}. \tag{22}$$

This geometry is thus singular at $r = 0$ for all $M \neq -1$ and it approaches anti-de Sitter space asymptotically. This black hole has the topology $\Re^2 \times S_3$.

3.3 Charges for the topological black hole

As we saw in the last section, global conserved charges can be found in a simple form in a Chern-Simons theory provided one couples an Abelian gauge field adding a term to the action of the form $b_\wedge g_{ab} F^a{}_\wedge F^b$. It turns out that in the context of five dimensional supergravity, this Abelian field is automatically present [16]. Indeed, supersymmetry requires an Abelian field b coupled to the gravitational variables by the term $b_\wedge R^{AB}{}_\wedge R_{AB}$, where R^{AB} is the anti-de Sitter curvature.

We then consider the topological black holes as solutions to the Chern-Simons and compute their mass and angular momentum as explained above. [Angular momentum is added by using a different Killing vector to perform the identifications. See [5] and [6] for more details.] The mass M and angular momentum J of the black hole embedded in this supergravity theory are,

$$M = \frac{2r_+ r_-}{l^2}, \qquad J = \frac{r_+^2 + r_-^2}{l}. \tag{23}$$

In the same way one can associate a semiclassical entropy to the black hole which is given by

$$S = 4\pi\, r_-. \tag{24}$$

This result is rather surprising because it does not give an entropy proportional to the area of S_1 $(2\pi r_+)$. A similar phenomena has been reported by Carlip et al [20]. The topological black hole thermodynamics in the context of standard general relativity has been analysed in [21].

The entropy given in (24) satisfies the first law,

$$\delta M = T\delta S + \Omega \delta J, \tag{25}$$

where M and J are given in (23) and $T = (r_+^2 - r_-^2)/(2\pi r_+ l^2)$, $\Omega = r_-/l r_+$.

Acknowledgments

During this work I have benefited from many discussions with Andy Gomberoff, Marc Henneaux, Cristián Martínez, Peter Peldán, Claudio Teitelboim, Ricardo Troncoso and Jorge Zanelli. This work was partially supported by the grant # 1970150 from FONDECYT (Chile), and institutional support by a group of Chilean companies (Empresas Cmpc, Cge, Copec, Codelco, Minera La Escondida, Novagas, Enersis, Business Design and Xerox Chile).

References

1. S. Aminneborg, I. Bengtsson, S. Holst and P. Peldan, *Class. Quant. Grav.* **13**, 2707 (1996).
2. M. Bañados, M. Henneaux, C. Teitelboim and Zanelli, *Phys. Rev.* **D48**, 1506 (1993).
3. K. Sfetsos and K. Skenderis, *Nucl. Phys.* **B517**, 179 (1998); J. Maldacena, *Adv. Theor. Math. Phys.* **2**, 231 (1998); A. Strominger, *J. High Energy Phys.* **02**, 009 (1998).
4. S. Holst and P. Peldan, *Class. Quant. Grav.* **14**, 3433 (1997).
5. M. Bañados, *Phys. Rev.* **D57**, 1068 (1998).
6. M. Bañados, A. Gomberoff and C. Martínez, in preparation.
7. M. Bañados, L.J. Garay and M. Henneaux, *Phys. Rev.* **D53**, R593 (1996).
8. M. Bañados, L.J. Garay and M. Henneaux, *Nucl. Phys.* **B476**, 611 (1996).
9. V.P. Nair and J. Schiff, *Phys. Lett.* **B246**, 423 (1990); *Nucl. Phys.* **B371**, 329 (1992).
10. A. Losev, G. Moore, N. Nekrasov and S. Shatashvili, *Nucl. Phys. Proc. Suppl.* **46** 130 (1996).
11. G. Moore and N. Seiberg, *Phys. Lett.* **B220**, 422 (1989); S. Elitzur, G. Moore, A. Schwimmer and N. Seiberg *Nucl. Phys.* **B326**, 108 (1989)
12. E. Witten, *Commun. Math. Phys.* **92**, 455 (1984).
13. A. P. Balachandran, G. Bimonti, K.S. Gupta, A. Stern, *Int. Jour. Mod. Phys.* **A7**, 4655 (1992); M. Bañados, *Phys. Rev.* **D52**, 5816 (1995).
14. D. Lovelock, *J. Math. Phys.* **12**, 498 (1971).
15. C. Teitelboim and J. Zanelli, *Class. & Quant. Grav.* **4**, L125 (1987) and in *Constraint Theory and Relativistic Dynamics*, edited by G. Longhi and L. Lussana, (World Scientific, Singapore, 1987).
16. A. H. Chamseddine, *Nucl. Phys.* **B346**, 213 (1990).
17. R. Troncoso and J. Zanelli, *Phys. Rev.* **D 58**, R101703 (1998).

18. M. Bañados, C. Teitelboim and J.Zanelli, *Phys. Rev.* **D49**, 975 (1994).
19. M. Bañados, R. Troncoso and J. Zanelli, *Phys. Rev.* **D54**, 2605 (1996).
20. S. Carlip, J. Gegenberg, R.B.Mann, *Phys. Rev.* **D51**, 6854 (1995).
21. J. D. E. Creighton and R.B. Mann, *Phys. Rev.* **D58**, 024013 (1998).

WORMHOLES ON THE WORLD VOLUME: BORN-INFELD PARTICLES AND DIRICHLET P-BRANES

G. W. GIBBONS
D.A.M.T.P.,
Cambridge University,
Silver Street,
Cambridge CB3 9EW,
U.K.

I describe some recent work in which classical solutions of Dirac-Born-Infeld theory may be used to throw light on some properties of M-theory. The sources of Born-Infeld theory are the ends of strings ending on the world volume. Equivalently the fundamental string may be regarded as merely a thin and extended piece of the world volume.

1 Introduction

In the lecture I described some recent work I had been doing over a period of time assisted by my research student Dean Rasheed and with some initial assistance from Robert Bartnik. Some related ideas had been discussed in an unpublished note of Douglas, Schwarz and Lowe [1]. After the lecture I was informed by Curt Callan that he and Maldacena had also been thinking along the same lines. Their work is to be found in [3] and my own in [2]. What follows is a slightly extended and expanded version of what I said in Santiago. The bibliography below is mainly restricted to some relevant papers which appeared during the autumn after the lecture. For a full set of references to earlier work the reader is referred to [2,3,1].

As is well known, p-dimensional extended objects, 'p-branes' play a central role in establishing the various dualities between the the five superstring theories and eleven-dimensional supergravity theory and these in turn which have led to the conjecture that there exists a single over-arching structure, called 'M-theory', of which they may all be considered limiting cases. Whatever M-theory ultimately turns out to be, it is already clear that it is a theory containing p-branes.

Until recently p-branes have been treated as

- Soliton-like BPS solutions of SUGRA theories.

  or

- The ends of open superstrings satisfying $9-p$ Dirichlet and $p+1$ Neumann boundary conditions.

My intention is to consider the ***light brane approximation*** in which Newton's constant $G \to 0$ but the string tension α' remains finite. In terms of actions we have:

$$S = \frac{1}{g_s^2} S^{\text{bulk}}_{NS\otimes NS} + S^{\text{bulk}}_{R\otimes R} + \frac{1}{g_s} S^{\text{brane}}, \tag{1}$$

where S^{bulk} is an integral over 10-dimensional spacetime and S^{brane} is an integral over the $p+1$=dimensional $p+1$ dimensional world volume of the brane. ***Heavy branes*** correspond to the limit $g_s \to \infty$. ***Light branes*** correspond to the limit $g_s \to 0$. It is reasonable to ignore the fields generated by the motion of the brane and set the Ramond-Ramond fields to zero. We therefore consider a Dirichlet p-brane moving in flat $d+1$ dimensional Minkowski spacetime $\mathbb{E}^{d,1}$ with constant dilaton and vanishing Kalb-Ramond 3-form.

2 The Dirac-Born-Infeld action

For purely bosonic fields S^{brane} is then given by the Dirac-Born-Infeld action

$$-\int_{\Sigma_{p+1}} d^{p+1}x \sqrt{-\det(G_{\mu\nu} + F_{\mu\nu})} \tag{2}$$

where

$$G_{\mu\nu} = \partial_\mu Z^A \partial_\nu Z^B \eta_{AB} \tag{3}$$

is the pull back of the Minkowski metric η_{AB} to the world volume Σ_{p+1} using the embedding map $Z^A(x^\mu) : \Sigma_{p+1} \to \mathbb{E}^{d,1}$ and

$$F_{\mu\nu} = \partial_\mu A_\nu - \partial_\nu A_\mu \tag{4}$$

is the curvature or field strength of an abelian connection $A_\mu(x^\nu)$ defined over the world volume.

The Dirac-Born-Infeld action is invariant under the semi-direct product of

- world volume diffeomorphisms and
- abelian gauge transformations.

To fix the former we adopt *Static Gauge*, called by mathematicians the ***non-parametric representation*** :

$$Z^M = x^\mu, M = 0, 1 \ldots, p \tag{5}$$

$$Z^M = y^m, M = p+1, \ldots, d-p. \tag{6}$$

The transverse coordinates y^m behave like scalar fields on the world volume and the action becomes

$$-\int_{\Sigma_{p+1}} d^{p+1}x \sqrt{-\det(\eta_{\mu\nu} + \partial_\mu y^m \partial_\nu y^m + F_{\mu\nu})}. \tag{7}$$

Note that original manifest global Poincaré symmetry $E(d,1)$ has been reduced to a manifest $E(p,1) \times SO(d-p)$.

An important message of this work is that static gauge cannot usually be globally well defined and it generates spurious singularities if the world volume is topologically non-trivial. Even if Σ_{p+1} is topologically trivial, static gauge may still break down if the brane bends back on itself. Geometrically we have projected Σ_{p+1} onto a $p+1$ hyperplane in $\mathbb{E}^{d,1}$ and the y^m are the height functions. However the projection need not be one-one.

Static gauge makes apparent that there are two useful consistent truncations

- $y^m = 0$ which is pure Born-Infeld theory[4,5] in $\mathbb{E}^{p,1}$

  and

- $F_{\mu\nu} = 0$ which corresponds to Dirac's theory[6] of minimal timelike submanifolds of $\mathbb{E}^{d,1}$.

The basic time independent solutions in these two cases are the BIon and the catenoid respectively. We shall see that there are two one-parameter families of solutions interpolating between them, rather analogous to the one parameter family of Reissner- Nordrstrom black holes. An internal Harrison like $SO(1,1)$ boost symmetry moves us along the two families, one of which is 'sub-extreme' and the other of which is 'super-extreme'. The two families are separated by an extreme BPS type solution.

3 BIons

Let's start with Born-Infeld theory and consider the case $p = 3$. Similar results hold if $p \neq 3$. The original aim of this theory was to construct classical finite energy pointlike solutions representing charged particles. It is these that I call 'BIons' and their study 'BIonics'.

For time independent pure electric solutions the lagrangian reduces to

$$L = -\sqrt{1 - \mathbf{E}^2} + 1. \tag{8}$$

where $\mathbf{E} = -\nabla\phi$ is the electric field. Thus the electric induction is

$$\mathbf{D} = \frac{\partial L}{\partial \mathbf{E}} = \frac{\mathbf{E}}{\sqrt{1-\mathbf{E}^2}} \tag{9}$$

and thus

$$\mathbf{E} = \frac{\mathbf{D}}{\sqrt{1+\mathbf{D}^2}}. \tag{10}$$

Now if

$$\nabla \cdot \mathbf{D} = 4\pi e \delta(\mathbf{x}), \tag{11}$$

$$\mathbf{E} = \frac{e\hat{\mathbf{r}}}{\sqrt{e^2+r^4}}. \tag{12}$$

Clearly while the induction $\mathbf{D}$ diverges at the origin the electric field remains bounded and attains unit magnitude at the origin. In other words the slope of the potential is 45 degrees at the origin.

The energy density is $T_{00} = \mathbf{E}\cdot\mathbf{D} - L$ and it is easy to see that the total energy is finite.

It is important to realize the difference, not widely understood, between 'BIons' and conventional 'solitons'. Originally Born-Infeld theory was intended as a 'unitary' theory of electromagnetism. In modern terms such a theory would be one in which the classical electron is represented by an everywhere non-singular finite energy of the source free non-linear equations of motion. In such theories the particle equations of motion follow from the equations of motion of the fields without having to be postulated separately. As such the theory was a failure because

- The BIon solutions have *sources* .
- The solutions are still *singular* at the location of the source.

  and

- One must impose *boundary conditions* on the singularities in order to obtain the equations of motion. For a recent discussion of this point see[7].

Nevertheless 'BIonic' solutions of field theories frequently have a sensible physical interpretation (cf. point defects in liquid crystals). To illustrate the point we consider briefly Born-Infeld electrostatics. Solutions of the equation of motion

$$\nabla \cdot \frac{\nabla\phi}{\sqrt{1-|\nabla\phi|^2}} \tag{13}$$

may be interpreted as spacelike maximal hypersurfaces $t = \phi(\mathbf{x})$ in an auxiliary $p+1$ Minkowski spacetime with coordinates $(t, \mathbf{x})$. This allows one to use geometrical techniques from general relativity and p.d.e. theory to discuss the existence and uniqueness of solutions. More significantly it allows us to construct new solutions. For example boosting the trivial solution with velocity $v = E$ in the z direction gives rise to a uniform electric field $\phi = -Ez$. The maximal field strength in Born-Infeld theory corresponds to the maximum velocity in special relativity.

One may also boost the BIon solution to give a *static* solution representing a charged particle at rest in an asymptotically uniform electrostatic field E. This sounds paradoxical but it is not. The point is that the solution does not satisfy the correct boundary conditions at the particle centre to be force free. It is pinned by a force $F = eE$ given by

$$F_i = \int T_{ij} d\sigma_j \tag{14}$$

where the integral is taken over a sphere surrounding the Bion.

Static solutions which *do* satisfy the force free solution can be also found. Thus if $\mathfrak{p}(x)$ is the Weierstrass elliptic function with $g_3 = 0$ and $g_2 = 4$ then

$$\mathfrak{p}(x)\mathfrak{p}(y)\mathfrak{p}(z) = \mathfrak{p}(\phi) \tag{15}$$

we get a BIon crystal of NaCl type. In this case the forces on the BIons cancel by symmetry. In general one may apply comparative statics and the virial theorem to obtain some striking analogues of the results in black hole theory. If $\mathbf{F}^a$ is the force on the a'th BIon which has position $\mathbf{x}_a$, charge e_a and electrostatic potential Φ^a one has the 'second law'

$$dM = \Phi^a de_a + \mathbf{F}^a \cdot d\mathbf{x}_a \tag{16}$$

and the Smarr-Virial relation:

$$M = \frac{1}{3}\mathbf{F}^a \cdot \mathbf{x}_a + \frac{2}{3} e_a \phi^a. \tag{17}$$

Here M is the total energy and there is a sum over the BIon index a.

4 Catenoids

Consider one transverse coordinate y. The lagrangian now becomes

$$L = -\sqrt{1 + |\nabla y|^2} + 1. \tag{18}$$

If $p = 3$ we soon find that a spherically symmetric solution satisfies

$$\partial_r y = \pm \frac{c}{\sqrt{r^4 - c^2}} \tag{19}$$

The solution breaks down at $r = \sqrt{c}$ because of a breakdown of static gauge. In fact the spatial part of the world volume (i.e. the p-brane) consists of two copies of the solution for $r > \sqrt{c}$ joined by a minimal throat. In other words, the solution has the geometry of the Einstein-Rosen throats familiar in Black Hole theory. (In fact the Einstein Rosen throats, i.e. the constant time surfaces of static black holes or of self-gravitating p-branes, are minimal submanifolds).

Near infinity the catenoid looks like two parallel p-planes situated a finite distance Y apart. Callan and Maldacena[13] have suggested that one should regard this as a D-brane-anti-D-brane configuration though it is a single connected surface. The catenoid is unstable in that it one can find a deformation which lowers the total volume. For that reason it was suggested by them that it should be thought of as some sort of sphaleron. It is interesting to note that subsequent to, and independent of, their discussion there appeared a paper [8] in which the it was shown, using the fact that that the Hessian (i.e second variation) of the Dirac energy is

$$\int \sqrt{g} d^p x f \left(-\nabla_g^2 - K_{ij} K^{ij} \right) f. \tag{20}$$

where K_{ij} is the second fundamental form of the hypersurface, that quite generally any complete minimal hypersurface of $\mathbb{E}^{p+1}$ with more than one end admits bounded harmonic functions and thus cannot be a true minimum of the energy.

5 Charged Catenoids

In general the relevant lagrangian is

$$L = -\sqrt{1 - |\nabla\phi|^2 + |\nabla y|^2 + (\nabla y \cdot \nabla\phi)^2 - (\nabla\phi)^2(\nabla y)^2} \tag{21}$$

This is manifestly invariant under generalized Harrison transformations consisting of boosts in the $\phi - y$ plane. Starting from the BIon or the catenoid we obtain the two one parameter families mentioned above. Note that the super-extreme solutions have a singular source on the world volume while the sub-extreme solutions are perfectly regular and have no source on the world volume. Starting with the catenoid and charging it up gives a narrower and narrower and longer and longer throat. Starting with the BIon and adding the scalar gives a bigger and bigger spike. The interesting question is what about the limiting case?

6 The BPS solution

It is a simple task to verify that taking

$$\phi = \pm y = H \tag{22}$$

where H is an arbitrary harmonic function will solve the equations. If

$$H = \sum_a \frac{c_a}{|\mathbf{x} - \mathbf{x}_a|}, \tag{23}$$

we get a superposition of arbitrarily many infintely spiky solutions. One may verify that these solutions are supersymmetric and indeed they satisfy the effective equations of motion of the superstring to all orders [11]. We shall see why shortly. In the mean time we point out that the obvious natural interpretation of these solutions is that they represent infinitely long fundamental strings ending on a D-brane as first envisaged by Strominger and by Townsend. We can now see clearly where the source for the BIon comes from. It is carried by the string. Indeed one may check that the charge carried by the string equals that carried by the BIon using the fact the the coupling to the Neveu-Scharz field $B_{\mu\nu}$ in the Dirac-Born-Infeld action is obtained by the replacement

$$F_{\mu\nu} \rightarrow \mathcal{F}_{\mu\nu} = F_{\mu\nu} - B_{\mu\nu}. \tag{24}$$

Note that if $c = 1$ the D-brane spike has height L at a distance $\frac{1}{L}$ from the source. The paper by Callan and Maldacena[3], see also [9] gives more detailed evidence for this viewpoint by showing that, by being careful about factors, the energy of a length L of string agrees with the world volume energy of the fields outside a radius $\frac{1}{L}$.

7 Electric-magnetic duality and the inclusion of magnetic fields

We have $\mathbf{H} = -\frac{\partial L}{\partial \mathbf{B}} = -\nabla\chi$ where χ is the magnetostatic potential. Let $\Phi^A = (y, \phi, \chi)$ be coordinates in an auxilliary Minkowski spacetime $\mathbb{E}^{1,2}$ (with two negative signs).

By means of a suitable Legendre transformation one may obtain an effective action from which to deduce the equations of motion. It is

$$\sqrt{\det(\nabla\Phi^A \cdot \nabla\Phi^B - \eta^{AB})}. \tag{25}$$

This is manifestly invariant under $SO(2,1) \supset SO(2)$. The $SO(2)$ subgroup of rotations of ϕ into χ is of course just electric-magnetic duality rotations. It is well known that Born-Infeld theory has this symmetry. In fact its existence may be traced back to the basic S-duality of non-perturbative string theory. Acting on the solutions with it we can obtain magnetically charged BIons attached to D-strings. In fact classically there is an entire circle of dyonic BPS solutions but of course quantization breaks down $SL(2,\mathbb{R})$ to $SL(2,\mathbb{Z})$.

8 Abelian Bogomol'nyi Monopoles

Setting $\phi = 0$ we find magnetic solutions with $\mathbf{B} = \mathbf{H}$. Thinking of y as a Higgs field we recognize the equations

$$\nabla y = \pm \mathbf{B} \tag{26}$$

as the abelian Bogomol'nyi equations of Yang-Mills theory, valid in the limit that the mass m_W of the vector bosons goes to infinity. This is consistent with Witten's ideas about nearby Dirichlet-branes. If two branes are well separated one has a gauge group $U(1) \times U(1)$, one of the factors corresponding to the centre of mass motion. If they are coincident one expects symmetry enhancement to $U(2)$. Taking out a $U(1)$ factor corresponding to the centre of mass the world volume gauge group is $SU(2)$. The distance Y between the branes is supposed to be proportional to m_W. The abelian Born-Infeld theory does indeed seem able to capture the physics of the large vector-boson mass limit.

9 Dimensional Reduction and SUSY

It is rather convenient to obtain the Dirac-Born-Infeld lagrangian in static gauge by dimensionally reducing the pure Born-Infeld lagrangian

$$-\sqrt{-\det(\eta_{AB} + F_{AB})} \tag{27}$$

from ten dimensions to $p+1$ dimensions. One sets

$$A_A = (A_m(x), A_\mu(x)) \tag{28}$$

and identifies the transverse components A_m of the gauge connection one-form A_A with the transverse coordinates y^m of the $p+1$ brane. At lowest order one may use the supersymmetry transformations of 1-dimensional SUSY (abelian) Yang-Mills. Thus SUSY requires the existence of a sixteen component Majorana-Weyl Killing spinor ϵ such that

$$F_{AB}\gamma^A\gamma^B\epsilon = 0. \tag{29}$$

In the electric case, our ansatz is

$$F_{5i} = F_{0i}, \tag{30}$$

so the BPS condition requires $(\gamma^0 + \gamma^5)\epsilon = 0$, which has eight real solutions. The self dual solutions are also easily seen to be BPS. One may check that both continue to admit killing spinors when the full non-linear supersymmetry transformations have been taken into account [18]. Moreover, in the electric case it has been argued that the solution gives an exact boundary conformal field theory[11]. This is almost obvious because of the lightlike nature of the the electric ansatz. All contractions involving F_{AB} must vanish.

The easily verified fact that the abelian anti-self-duality equations are sufficient conditions for solutions the Born-Infeld equations leads to an interesting and useful relation to minimal 2-surfaces in $\mathbb{E}^4$. One assumes that A_μ depends only on $z = x^1 + ix^2$. Setting $A_3 + iA_4 = w = x_3 + ix_4$ one then easily calculates that $F_{\mu\nu} = - \star F_{\mu\nu}$ reduces to the Cauchy-Riemann equations, i.e. to the condition that w is a locally holomorphic function of z. In this way one may obtain a variety of interesting minimal surfaces which can in fact be regarded as exact solutions of M-theory. Thus if

$$wz = c, \tag{31}$$

and $c \neq 0$, we obtain a smooth connected 2-brane with topology $\mathbb{C}^\star \equiv \mathbb{C} \setminus 0$ looking like two 2-planes connected by a throat. If $c = 0$ this degenerates to two 2-planes, $z = 0$ and $w = 0$ intersecting at a point. Constructions of minimal surfaces in $\mathbb{E}^4$ using holomorphic embeddings were pioneered by Kommerell in 1911 [12] so it seems reasonable to refer to them as Kommerell solutions. The reader is referred [13] for an account of recent applications to gauge theory.

10 Calibrated geometries

The holomorphic solutions of the last section are in fact a special case of a more general class of solutions in which the familiar Wirtinger's inequality for the area, or more generally the p-volume, of holomorphically embedded p-cycles is replaced by a more general inequality. The basic idea is to replace a suitable power of the Kähler form by some other closed p-form. The form is called by Harvey and Lawson [14] a calibrating p-form Using their work one see that the Dirac-Born-Infeld equations have a very rich set of solutions. Here is not the place to discuss them and their applications to gauge theory in detail. I will simply remark that one encounters various kinds of topological defects on the world volume, such as vortices and global monopoles. These latter may be relevant to discussions of the non-abelian monopoles in the limit that the gauge field decouples. Of special interest are the BPS solutions and their Bogomol'nyi bounds. It turns out that there is a close connection between this, kappa symmetry, and the calibration condition of Harvey and Lawson[14]. In fact he calibration condition turns out to be the condition for supersymmetry[15].

11 Non-Abelian Born-Infeld

It is widely believed that when a number of D-branes coincide there is symmetry enhancement. The current most popular suggestion for the relevant generalization of the Born-Infeld action is that of Tseytlin

$$-\mathrm{Str}\sqrt{-\det(\eta_{AB}+F_{AB})} \tag{32}$$

where F_{AB} is in the adjoint representation of the gauge group G and Str denotes the symmetrized trace of any product of matrices in the adjoint representation that it precedes. The Tseytlin action has the property, [16,9] that solutions of the non-abelian Bogomol'nyi equations are also solutions. In the case of $SU(2)$ one may even establish a generalized Bogomol'nyi bound [16]. Very recently[17] BPS bounds have also been established in a rather different way using an apparently different energy functional.

12 Conclusion

A striking aspect of the work reported above is the way in which Born-Infeld theory has finally found a home on the world-volume and its mysterious sources have been shown to be just the ends of strings extending into higher dimensions. Even more striking is the way in which the fundamental string solution emerges as a limiting case of M-theory solutions. It seems to reinforce the widely held viewpoint that in the ultimate formulation of the theory, strings as such may have no fundamental role to play and may indeed appear only as effective excitations. However, as always, it is worth exercising some caution. After all, who would have thought a few years ago that Born-Infeld theory and Dirac's doomed attempt to construct an extended model of the electron, long since relegated to the dustbin of history and condemned as a last nostalgic gasp at the fag-end of the classical world-picture should re-emerge at the cutting edge of post modernist physics?

There will no doubt be many more fag-ends and even a few cigars before the final story is told. In the meantime it is my pleasant duty to thank Claudio and Jorge, so ably assisted by the Chilean Air Force, for organizing such a wonderful conference and making our stay in Chile and Antartica so memorable.

References

1. Mike Douglas, David Lowe and John Schwartz, unpublished notes and private communication
2. G. W. Gibbons, Born-Infeld particles and Dirichlet p-branes, *Nucl. Phys.* **B514** (1998) 603
3. C. Callan and J. Maldacena, Brane death and dynamics from the Born-Infeld action, *Nucl. Phys.* **B513** (1998) 198
4. M. Born and L. Infeld, Foundations of The New Field Theory, *Proc. Roy. Soc.*, **A144** (1935) 425
5. M. Born, Théorie non-linéare du champ électromagnétique *Ann. Inst. Poincaré*, **7** (1939) 155-264
6. P. A. M. Dirac, An extensible model of the electron, *Proc. Roy. Soc. Lond.* **A268** (1962) 57-67
7. A. A. Chernitskii, Non-linear electrodynamics with singularities (Modernized Born-Infeld Electrodynamics), *Helv. Phys .Acta* **71** (1998) 274
8. H.-D. Cao, Y. Shen and S. Zhu, The structure of stable minimal hypersurfaces in $\mathbb{R}^{n+1}$, dg-ga/9709001
9. A. Hashimoto, The Shape of Branes Pulled by Strings, *Phys. Rev.* **D57** (1998) 6441
10. S. Lee, A. Peet and L. Thorlacius, Brane-waves and strings, *Nucl. Phys.* **B514** (1998) 161
11. L. Thorlacius, Born-Infeld String as a Boundary Conformal Field Theory, *Phys. Rev. Lett.* **80** (1998) 1588
12. K. Kommerell, Strahlensysteme und Minimalflächen, *Math. Ann.* **70** (1911) 143-160
13. A. Mikhailov, BPS States and Minimal Surfaces, *Nucl. Phys.* **B533** (1998) 243
14. R. Harvey and H. B. Lawson, Calibrated Geometries *Acta Mathematica* **148** (1982) 47-157
15. G.W. Gibbons and G.P. Papadopoulos, Calibrations and Supersymmetry for Born-Infeld, manuscript in preparation
16. G. W. Gibbons and A. Tseytlin, unpublished
17. J.P. Gauntlett, J. Gomis and P.K. Townsend, Bounds for Worldvolume Branes, *J. High Energy Phys.* **01** (1998) 003
18. P. Howe, N. Lambert and P. West, The self-dual string soliton, *Nucl. Phys.* **B515** (1998) 203

EVAPORATION OF PRIMORDIAL BLACK HOLES

S.W. HAWKING
Department of Applied Mathematics and Theoretical Physics,
Silver Street, Cambridge, CB3 9EW, UK
E-mail: hawking@damtp.cam.ac.uk

The usual explanation of the isotropy of the universe is that inflation would have smoothed out any inhomogeneities. However, if the universe was initially fractal or in a foam like state, an overall inflation would have left it in the same state. I suggest that the universe did indeed begin with a tangled web of wormholes connecting pairs of black holes but that the inflationary expansion was unstable: wormholes that are slightly smaller correspond to black holes that are hotter than the cosmological background and evaporate away. This picture is supported by calculations with Raphael Bousso of the evaporation of primordial black holes in the s-wave and large N approximations.

1 Introduction

The universe around us is remarkably homogeneous and isotropic on a large scale. Indeed it took 27 years after the discovery of the microwave background before we were able to find any structure in it and even then, the variations were only 1 in 10^5. The most attractive explanation of this amazing isotropy is inflation. The idea is the universe had a period of exponential expansion in the very early stages. The expansion is supposed to have stretched out any lumps and bumps in the early universe so that now they look flat locally. If the initial metric in the early universe were smooth on small scales, a uniform expansion would indeed have the effect of making the universe locally homogeneous and isotropic. On the other hand, if the initial state of the universe were a fractal with structure on smaller and smaller scales, then blowing it up with inflation would lead to a universe that was still inhomogeneous and anisotropic.

Because Newton's constant has dimensions of length to the minus two one would expect quantum fluctuations of the metric to be larger, the smaller the length scale. Thus one might expect spacetime to have a fractal or foam like structure on scales of the Planck length or less. The amount of inflation needed to explain the present flatness of the universe is at least a factor of 10^{28}, followed by a further expansion by the same factor. That would have blown the Planck length in the early universe up to bigger than the galaxy now. So why is the metric in the galaxy smooth and almost flat. What happened to all that spacetime foam?

I shall endeavour to answer these questions in this talk which is based on joint work with my student Raphael Bousso [1]. I shall argue that it is natural to expect the universe began its inflation in a foam like state, with all possible topologies for the space sections. The simplest topology is the three sphere, but one can also have three spheres with wormholes or handles connecting different regions. As they expand such three geometries represent expanding universes with pairs of black holes at the ends of the wormholes. One could either say that this process represented the inflation of spacetime foam, or that it was the pair creation of black holes by a cosmological constant. In my opinion these two descriptions are equivalent and it is a waste of time to argue which is correct. The physical effects are the same.

In order to explain the present flatness of the universe it seems necessary to postulate that the inflationary period went back to when the universe was the Planck size. In that situation, as I will show, black holes would have been pair created in abundance. Equivalently, one could say that the probability of a space section with n handles would not go down rapidly as n increases. So why doesn't inflation blow up these space sections to produce a universe that is full of macroscopic black holes? I shall argue that while uniform inflation is possible it is quantum mechanically unstable. What happens is, that at early times there is no clear distinction between the three sphere and the handles, or between black hole and cosmological horizons. They have similar sizes and temperatures initially. But as the universe inflates generally some will increase in size less than others. One might expect such smaller handles to be hotter than the larger ones and so to lose energy to them in thermal radiation. This would further increase the difference in size between the smaller and larger handles. Eventually smaller handles that do not have a gauge field through them could shrink back down to Planck size. They would then either pinch off, or just disappear into the sea of Planck scale fluctuations of the metric that presumably occurs even at the present time. Handles with a net gauge flux could stabilize as near extreme charged black holes. However, their initial probability will be reduced relative to that of neutral handles by the scale independent action of the gauge field. This means that their density will be greatly diluted by inflation that blows sub Planck scales up to the size of the galaxy today.

To discuss the evaporation of handles or black holes in the early univers, one really needs to know the effective action, to one loop at least. One might avoid problems with divergences by working with super gravity theories that are known to be finite at one and two loops, but the one loop effective action for a general background is not known. Instead Raphael and I used a different approximation. We assumed that the evaporation was dominated by spherically symmetric field configurations and by a large number of scalar fields, or effective scalars. This meant we could dimensionally reduce to two dimensions and use the large n expansion. There was a great industry in such two dimensional models of black holes a few years ago. Funnily enough, no one seems to have applied them in the cosmological context. Or at least not to my knowledge.

2 Initial Conditions

To discuss the formation of black holes in the early universe one needs a theory of initial conditions. There are three main candidates:

1. **The Tunnelling Hypothesis**
2. **The No Boundary Proposal**
3. **The Pre-Big Bang Scenario**

I won't waste time on the tunneling idea. In my opinion it is dead in the water. Even its principal proponent Vilenkin uses instanton methods to calculate the quantum creation of cosmic strings. But if instantons are appropriate for gauge field defects, why not for gravitational defects like black hole, or even for the whole universe? But the use of instantons is effectively the no boundary proposal.

However, I reserve my real venom for the pre-big bang scenario. One would have hoped that by now Physicists would have absorbed Einstein's message that time is a coordinate label in a dynamic spacetime rather than a fixed background. So it may not continue backwards indefinitely, as those behind the pre-big bang model assume it must. They achieve this by joining on before the big bang another expanding solution that ends on a singularity. It is claimed that higher order string effects will somehow remove the singularities and enable the two solutions to join smoothly. However, this is only a pious hope as no mechanism has been found so far. This means that the pre-big bang model has no predictive power because it is not known what equations, if any, hold during the smoothed period. The model claims respectability because the pre and post big bang solutions are related by T duality. But this claim is spurious because if smoothing were possible, and I see no reason to believe it is, it would break T duality.

Having given my opinion of the alternative theories of initial conditions I now turn to what I consider the only viable candidate, the no boundary proposal. In the ultimate unified theory the no boundary proposal should probably be applied in 11 or 12 dimensions, but in this paper I shall stick to the four dimensions that we know exist.

3 The No Boundary Proposal

The basic idea is that the quantum state of the universe is determined by a path integral over compact manifolds without boundaries. If the manifolds are simply connected, which I shall assume, a co-dimension one surface Σ will divide a spacetime manifold M into two parts. One can then factorize the probability into the product of two wave functions. The wave functions are defined by path integrals, over metrics on the two half manifolds M^+ and M^- bounded by Σ with the given induced metric h_{ij}. The wave functions obey the Wheeler–De Witt equation and the momentum constraints as functionals of h_{ij}. However, I shall estimate the wave functions by saddle point approximations to the path integral rather than by solving the Wheeler–De Witt equation as this is the only way of implementing the no boundary condition.

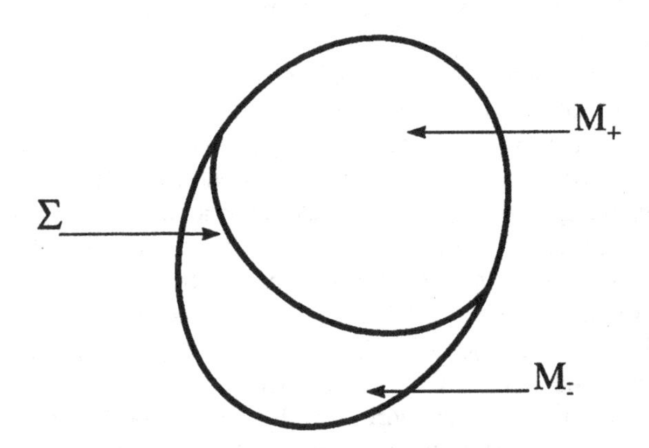

Wave Function of the Universe $\psi = \int_{M_+} d[g] e^{-I}$

Figure 1. The No Boundary Proposal

4 Inflation and the No Boundary Proposal

It seems reasonable to suppose that the early universe had a period of inflationary expansion driven by a scalar field χ with effective potential V. In regions in which χ is slowly varying one can neglect gradient terms. So the energy momentum tensor of χ, will be like a cosmological constant. For most of this talk I shall work with this effective cosmological constant Λ rather than the scalar field χ.

$$\Lambda = 8\pi V(\chi) \tag{1}$$

The approach to quantum cosmology that has been followed in the past is to examine the behavior of the wave function as a function of the overall scale a of the metric h_{ij} on the surface Σ. If the dependence on a was exponential this was interpreted as corresponding to an Euclidean spacetime while an oscillatory dependence on a was interpreted as corresponding to a Lorentzian spacetime. For example, in the case of Einstein gravity with a cosmological constant the path integral for the wave function of a three sphere of radius a will be dominated by an instanton

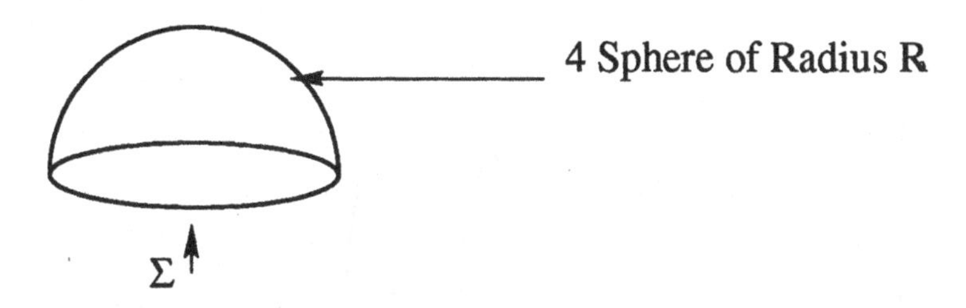

Figure 2. Instanton for the Wave Function $a < R_0$

This is part of a four sphere with radius $R_0 = \sqrt{3/\Lambda}$. For $a < R_0$ there will be a real Euclidean geometry M^+ bounded by the three sphere Σ of radius a. The wave function ψ will be 1 for $a = 0$ and will increase rapidly with a up to $a = R_0$. For $a > R_0$ there are no Euclidean solutions with the given boundary conditions. There are however two complex solutions.

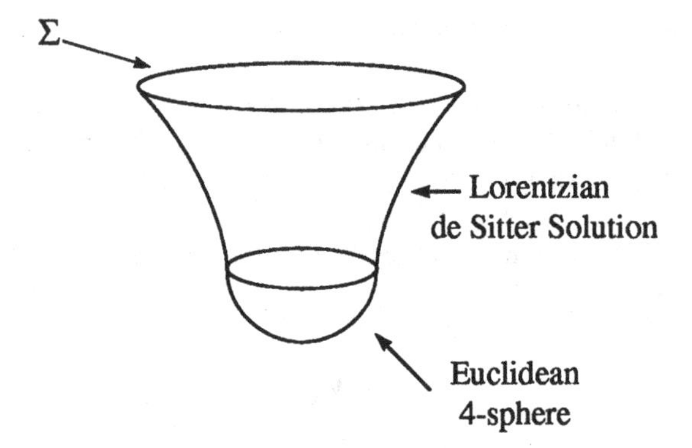

Figure 3. Complex Solution for $a > R_0$

Each of these complex solutions can be thought of as half the Euclidean four sphere joined to part of the Lorentzian De-Sitter solution. The real part of the action of these complex solutions is equal to the action of the Euclidean half four sphere and is the same for all values of a greater than R_0. On the other hand the imaginary part of the action comes from the Lorentzian De-Sitter part of the solution and depends on a. Thus the wave function for large a oscillates rapidly with constant amplitude, as a increases.

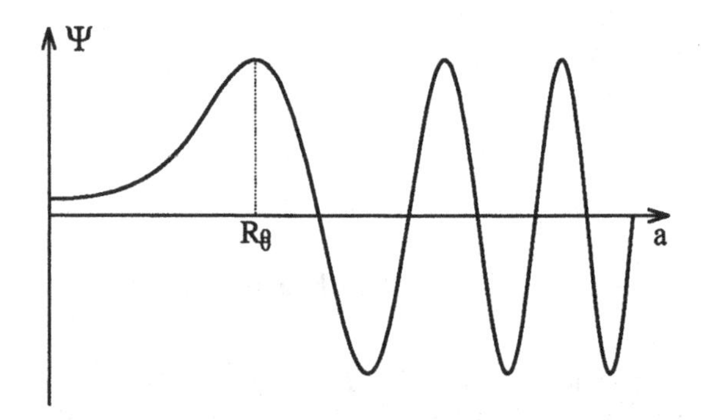

Figure 4. Wave Function in the Metric Representation

This shows the association between an oscillatory wave function and a Lorentzian spacetime. However, the distinction between exponential and oscillatory is not precise and does not identify how much of the wave function describes which physical situation. So it is not clear how to calculate the probability of Lorentzian geometries.

I therefore want to put forward a different approach which focuses on the defining characteristic of a Lorentzian geometry in the neighbourhood of Σ. This is that the induced metric h_{ij} on Σ should be real but the second fundamental form defined for Euclidean signature should be purely imaginary. The second fundamental form can be regarded as the derivative of the metric h_{ij} on Σ as Σ is moved in its normal direction in M. Thus requiring the second fundamental form to be purely imaginary means that h_{ij} has a real derivative with respect to the time coordinate $t = i\tau$ where τ is Euclidean time. This is the condition for a Lorentzian geometry in a neighbourhood of Σ.

$$Re(K^{ij}) = 0 \qquad (2)$$

The no boundary quantum state for the universe is normally described by a wave function ψ as a functional of the metric h_{ij} on Σ. However, one can transform to the momentum representation in which the wave function is a functional of the second fundamental form of Σ. The momentum representation is given by a Laplace transform of ψ with respect to the metric h_{ij} at each point of Σ.

5 Wave Function in the Momentum Representation

$$\psi(\pi^{ij}) = \int d[h_{ij}]\ \psi(h_{ij})\ exp[-\int \pi^{ij} h_{ij}] \qquad (3)$$

$$\pi^{ij} = \sqrt{h}(K^{ij} - h_{ij}K) \qquad (4)$$

For negative Euclidean second fundamental form the Laplace transform should converge. One could then analytically continue it to complex values of the second fundamental form. In this way, one can calculate the probability of an imaginary second fundamental form and so of a Lorentzian geometry near Σ. One can illustrate this with the De-Sitter model:

6 De-Sitter Model

One calculates the wave function for a three sphere Σ with uniform second fundamental form K^{ij}. There is a real Euclidean saddle point solution for all real values of K. It is part of the Euclidean four sphere of radius R_0. K large and positive, corresponds to a small segment of the four sphere around the north pole. The action of this segment is small. Thus the wave function ψ is one for large positive K. As K decreases the segment of the four sphere increases. When $K = 0$ it is half the four sphere with action $-3\pi/2\Lambda$. For K large and negative, it is almost the whole four sphere.

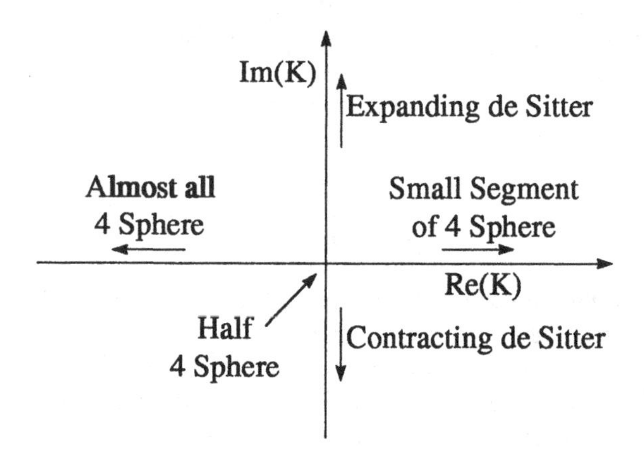

On Imaginary K $|\psi| = exp(3\pi/2\Lambda)$

Figure 5. Variation of ψ in K plane

Having calculated the wave function for real K one can now analytically continue to complex values. Up the imaginary K axis the amplitude of the wave function will remain at the value for $K = 0$ which is $e^{3\pi/2\Lambda}$. But the phase of the wave function will vary rapidly with the imaginary part of K. The wave function for positive imaginary K will be given by just one of the two complex solutions we had before. It is the one that consists of the half Euclidean four sphere joined at the time of minimum radius to an expanding De-Sitter solution. Thus this approach separates the expanding and contracting phases of the De-Sitter universe, which are mixed together when one looks at the wave function in the h_{ij} representation. It also makes contact with the with the tunneling proposal for the wave function of the universe. The condition that the second fundamental form corresponds to a Lorentzian expansion is the outgoing wave condition. This is used in the tunneling proposal to select solutions of the Wheeler–De Witt equation that correspond to a universe quantum tunneling from nothing. But the advocates of the tunneling proposal have been misled by the fact that the action of the Euclidean four sphere is negative. Thinking that the creation of the universe should be suppressed, they have argued that the probability of a universe appearing from nothing is e^{+I} rather than e^{-I}. But that would be inconsistent with the effect of perturbations and topological fluctuations like black holes.

7 Primordial Black Holes

To get back to black holes, one would like to calculate the probability for a Lorentzian geometry on a surface Σ with n handles. This would represent an expanding universe with n pairs of black holes that inflated from spacetime foam. It seems reasonable to suppose the probability of n handles is roughly the nth power of the probability of a single handle with appropriate phase space factors. Thus it is sufficient to consider the relative probabilities for zero and one handles. The most probable configurations will be spherically symmetric in both cases. I shall therefore restrict myself to spherical fields.

The zero handle surfaces correspond to the Lorentzian De-Sitter solution while the one handle surfaces correspond to Schwarzschild De-Sitter.

8 Schwarzschild-de-Sitter

$$ds^2 = -U(r)dt^2 + U(r)^{-1}dr^2 + r^2 d\Omega_2^2$$

$$U(r) = 1 - \frac{2M}{r} - \frac{1}{3\Lambda r^2}$$

$$0 < M \leq \frac{1}{3\sqrt{\Lambda}} \tag{5}$$

This represents a pair of black holes in a De-Sitter background. The mass, M, of the black holes can be in the range from zero up to a maximum value of $1/3\sqrt{\Lambda}$. For mass less than the maximum value the surface gravity of the black hole horizon is greater than that of the cosmological horizon. This means that if one tries to turn the Schwarzschild De-Sitter solution into a compact Euclidean instanton one gets a conical singularity either on the black hole horizon, or on the cosmological horizon. For this reason it has been thought that black holes could spontaneously nucleate in a De-Sitter background only if they had the maximum mass. In this case the Schwarzschild De-Sitter solution degenerates into the Narai solution, in which the two horizons have the same area and surface gravity. Thus a compact Euclidean instanton is possible without conical singularities. However, conical singularities on Σ are OK if they correspond to components of h_{ij} that are measured. For example, if one wants the probability of a handle with a two sphere cross section h of area A one can impose the Lorentzian condition that the real part of the second fundamental form vanishes everywhere on Σ except h. One can not specify the second fundamental form on h because one is prescribing the metric there. On the other hand one can impose the Lorentzian condition that the real part of the second fundamental form is zero everywhere else on Σ. This allows one to find a saddle point solution bounded by a surface Σ with a handle of area A for any area up to the maximum $4\pi/\Lambda$.

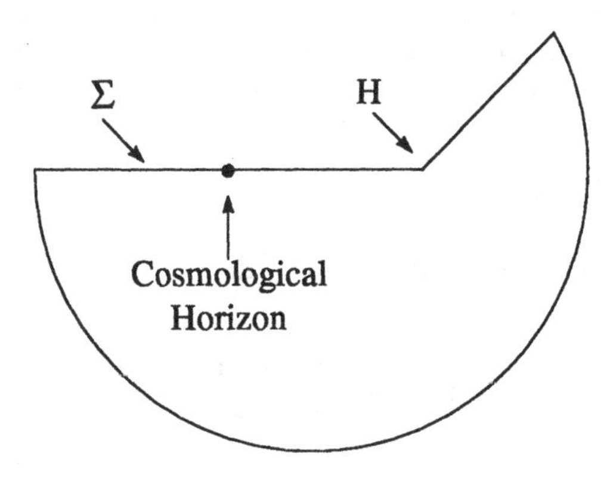

Figure 6. Instanton for Handle of Area A

The Euclidean part of the saddle point metric has a boundary with zero second fundamental form everywhere except on h where it is a delta function. The action of the saddle point metric minus the action without handles gives the relative probability of having a handle of area A. This probability decreases as the area increases and is $e^{-\pi/\Lambda}$ for the maximum area case.

One can see from this that the probability of handles or black holes is small unless the effective cosmological constant is near the Planck value. However, to account for the present state of the universe it is necessary to suppose that the initial value of the scalar field corresponds to a high value of the potential V and a large effective cosmological constant. This would mean that the production of handles would not be significantly suppressed. One would expect the universe to begin its Lorentzian expansion with a tangled web of handles. As the universe expands the scalar field will run down to the minimum of the potential and the effective cosmological constant will decrease. This can cause the handles to expand and become macroscopic black holes. The black holes will send out thermal radiation and absorb radiation from the surrounding universe. One would expect that black hole horizons that are much smaller than the cosmological horizon would radiate more than they absorb and would shrink and disappear. However, it is not so obvious what happens to a black hole horizon that is almost the size of the cosmological horizon.

To investigate this fully one would need to calculate the one loop effective action for a general cosmological metric. This is too difficult, but Raphael Bousso and I have studied an approximation that is tractable and may give an idea what will happen. The most probable black holes are spherically symmetric and the dominant emission is scalars in the s-wave. We therefore restricted to the case in which all fields are spherically symmetric. One can then dimensionally reduce the classical action to get something similar to the CGHS model [2], but with an extra term.

9 Reduced Classical Action for Spherically Symmetric Fields

$$ds^2 = g_{\mu\nu}dx^\mu dx^\nu + e^{-2\phi}d\Omega^2 \tag{6}$$

$$S = \frac{1}{16\pi}\int d^2x\,(-g)^{1/2}\,e^{-2\phi}[R + 2(\nabla\phi)^2 + 2e^{2\phi} - 2\Lambda - \sum_i(\nabla f_i)^2] \tag{7}$$

The idea both in CGHS and in the present case is to take the number N of scalar fields f to be large. Their quantum fluctuations will therefore dominate the quantum fluctuations in ϕ and ρ which are neglected. In two dimensions the effective action of a field is determined by the trace anomaly up to terms that depend on boundary conditions. However, the trace anomaly for scalars with a coupling to the ϕ field is different to that for the minimally coupled scalars that CGHS use. Previous authors like Strominger and Trevedi who have considered the dimensional reduction of four dimensional fields have brushed this awkward fact under the carpet but we felt we should take it into account. We couldn't find it in the literature so we calculated it and wrote a short piece [3]. It was a very minor paper but it provoked six other papers on the same subject and a referee who was unhappy about some comments we made about CGHS.

10 Effective Action of Scalars Coupled to ϕ

$$W = -\frac{1}{48\pi}\int d^2x\sqrt{g}\,[\frac{R}{2}\frac{1}{\Box}R - 6(\nabla\phi)^2 R\frac{1}{\Box}R - 2\phi R] \tag{8}$$

One can incorporate the effective action of the scalar fields by introducing an auxiliary field Z that couples to the trace anomaly. One then gets equations of motion for ρ and ϕ that describe quantum cosmology in the s-wave and large N approximations.

11 Equations of Motion

$$\kappa \equiv 2N/3$$

$$-(1-\frac{\kappa}{4}e^{2\phi})\partial^2\phi \;+\; 2(\partial\phi)^2 \;+\; \frac{\kappa}{4}e^{2\phi}\partial^2 Z \;+\; e^{2\rho+2\phi}(\Lambda e^{-2\phi}-1) = 0 \tag{9}$$

$$(1-\frac{\kappa}{4}e^{2\phi})\partial^2\rho \;-\; \partial^2\phi \;+\; (\partial\phi)^2 \;+\; \Lambda e^{2\rho} = 0 \tag{10}$$

$$\partial^2 Z \;-\; 2\partial^2\rho = 0 \tag{11}$$

These equations seem too nasty to allow closed form solutions so Raphael and I used perturbation theory. We started with the Narai solution, the limit of Schwarzschild De-Sitter, in which the black hole and cosmological horizons have the same areas and temperatures. In this dimensional reduction that is represented by a solution in which ϕ is constant everywhere $g_{\mu\nu}$ corresponds to two dimensional De-Sitter space while the conformal factor ρ is independent of the spatial coordinate, and has a maximum on a surface of time symmetry. One can decompose perturbations about this solution in a Fourier series and obtain differential equations for the time development of each term. The lowest order term has two linearly independent solutions. The time symmetric mode corresponds to a perturbation that changes the Narai solution into Schwarzschild De-Sitter with nearly equal horizons. One might have expected that the smaller black hole horizon would be hotter than the cosmological horizon and would radiate more than it absorbed. This would increase the difference between the horizons and so on. However, to our surprise the equations showed the opposite behavior. The horizons started to become more equal. In other words the black holes anti-evaporate. One might wonder whether this behavior was the result of using scalars whose kinetic term is coupled to the field ϕ. For black holes in asymptotically flat space the use of the trace anomaly for such ϕ coupled scalars can give a negative rate of radiation for some choices of the regulator dependent box ϕ terms in the trace anomaly. But in the cosmological case we are considering the gradients of ϕ are small. This means we would have obtained anti evaporation of Schwarzschild De-Sitter even if we had used the trace anomaly for minimally coupled scalars, like CGHS did. With minimally coupled scalars black holes in asymptotically flat space give positive energy radiation.

The explanation of this unexpected result of anti evaporation must be that in these two dimensional models the initial data for ρ and ϕ determines the distribution of the quantum radiation through the constraint equations. It must be that the constraints determine that more radiation is heading towards the smaller horizon.

The other linearly independent solution of the perturbation equations corresponds to no change in the geometry at the time of minimum size of the universe. However, there is a disturbance in the quantum radiation that gives a net flux of energy from one half of the S1 cross S2 initial surface, to the other. As expected this produces a difference between the horizons that increases linearly with time.

We therefore have two independent solutions for the equations that govern the lowest Fourier mode of perturbations of the maximum area handle. The no boundary condition will select one linear combination of these solutions as being present in the universe. So the question is, do the differences between the horizons grow with time in this linear combination, or do they decrease? To answer this one has to solve the perturbation equations on the Narai background with the boundary condition that the perturbation is regular at the south pole of the Narai two sphere. The perturbation calculations are a bit messy, so I will leave them to Raphael Bousso to describe in another paper.

12 Conclusion

The net result of the equations is that at late times the difference between the horizons grows exponentially. Thus it seems that the equilibrium between black hole horizons of maximal size and cosmological horizons is unstable. As the tangled web of handles in the spacetime foam inflates some regions expand more than others. The handles that get left behind radiate and shrink back to the Planck size. This leaves a universe that is homogeneous and isotropic from the Planck scale up to the scale of the cosmological horizon. On smaller or larger scales the geometry may be very different but in the range we can observe it is smooth and almost flat. That is fortunate, because otherwise we wouldn't exist!

References

1. *'(Anti-) Evaporation of Schwarzschild-de-Sitter Black Holes'*, R. Bousso, S.W. Hawking, Phys. Rev. **D57** 2436 (1998) .
2. *'Evanescent Black Holes'*, C.G. Callan, S.B. Giddings, J.A. Harvey and A. Strominger, Phys. Rev. **D45** 1005 (1992).
3. *'Trace Anomaly of Hidden Dilation Coupled Scalars in Two Dimensions'*, R. Bousso, S.W. Hawking, Phys. Rev. **D56** 7788 (1997).

FERMION MODELS AND CHERN SIMONS THEORIES

FIDEL A. SCHAPOSNIK *
Departamento de Física, Universidad Nacional de La Plata
C.C. 67, 1900 La Plata, Argentina

I present in this talk some results on three dimensional fermionic systems and Chern-Simons models. In particular I describe a bosonization approach that allows to establish a mapping between fermionic currents onto equivalent bosonic ones and also discuss the calculation of the effective action for planar fermions at finite temperature.

1 Introduction

The existence of a connection between bosonic theories containing a Chern-Simons (CS) term and fermionic models in $2+1$ dimensions was discovered some ten years ago in works by Dzyaloshinski, Polyakov and Wiegmann [1]-[2]. In particular, Polyakov demonstrated in [2], by explicit computation of propagators, that a Chern Simons term in the gauge action of CP^1 fields $\vec{n}$ turns bosons into fermions and viceversa. More explicitely, the dressing of the $\vec{n}$ quanta by the gauge field with a CS dynamics turns them into Dirac fermions: one has at large momenta bosons that behave at small momenta as fermions. This Fermi-Bose transmutation prompted many investigations in connection with fractional statistics and anyon physics [3].

Even before these works, the Chern-Simons term was considered as an action for gauge fields by Schonfeld [4] and by Deser, Jackiw and Templeton [5]. This term has the peculiarity of not containing the metric tensor so that in this sense it is a topological object. Because of this, the partition function for a *pure* CS theory has to be a topological invariant which, as first proved by Schwartz [6] corresponds to the Ray-Singer torsion. After the work of Witten on Jones polynomials [7], pure CS theories become a powerful tool to define and study knot and link invariants in arbitrary 3 manifolds.

The study of 2+1 gauge theories containing a CS term coupled to fermions also led to very interesting results [5,8]. In particular, it was discovered that a CS term is induced due to radiative fermion corrections so that considering the effect of this term is uncontournable in $2+1$ fermion systems: even if it is not included from the begining, a parity-anomaly generates it at the gauge field effective action level.

---

*INVESTIGADOR CICBA, ARGENTINA

Exploiting all this facts we have developed, together with N. Bralić, C. Fosco, E. Fradkin, J.C. Le Guillou, M.V. Manías, E. Moreno C. Núñez and G. Rossini, a bosonization approach that allows to connect three dimensional purely fermionic theories with bosonic theories containing a CS term [9-15]. The method is based in dualization of fermionic systems [16-19] and a more extended version of this talk can be found in [20]. Related and unrelated approaches to $d > 2$ bosonization have been developed in [21-30].

Before entering into the subject, let us list our conventions and some useful formulae. We work in $2+1$ Euclidean space-time with Dirac matrices chosen as follows:

$$\gamma_1 = \sigma_1 \quad \gamma_2 = \sigma_2 \quad \gamma_3 = \sigma_3 \tag{1}$$

The Chern-Simons action for an Abelian gauge field A_μ is defined as

$$S_{CS}[A] = \frac{i}{4\pi} \int d^3x \epsilon_{\mu\nu\alpha} A_\mu \partial_\nu A_\alpha \tag{2}$$

In the non-Abelian case, we write the gauge field connection A_μ as

$$A_\mu \;=\; A_\mu^a \tau_a \tag{3}$$

with τ_a denoting hermitian generators of the Lie algebra of $SU(N)$ ($a = 1, \ldots, N^2 - 1$), verifying the relations

$$\tau_a^\dagger \;=\; \tau_a, \quad [\tau_a, \tau_b] \;=\; i f_{abc} \tau_c, \quad \mathrm{tr}(\tau_a \tau_b) = \frac{1}{2} \delta_{ab} \,, \tag{4}$$

with f_{abc} the totally antisymmetric structure constants. Then the CS action is defined as

$$S_{CS}[A] = \frac{i}{8\pi} \int d^3x \epsilon_{\mu\nu\alpha} tr(f_{\mu\nu} A_\alpha - \frac{2}{3} A_\mu A_\nu A_\alpha) \tag{5}$$

with

$$f_{\mu\nu} = \partial_\mu A_\nu - \partial_\nu A_\mu + [A_\mu, A_\nu] \tag{6}$$

In the non-Abelian case, even if one disregards surface terms the action is not invariant under "large" gauge transformations. Indeed, given a gauge group element U with winding number defined as

$$w(U) = \frac{1}{12\pi^2 N} tr \int d^3x \epsilon_{\mu\nu\alpha} U^{-1} \partial_\mu U U^{-1} \partial_\nu U U^{-1} \partial_\alpha U \tag{7}$$

the CS action changes as

$$S_{CS}[A] \rightarrow S_{CS}[A^U] = S_{CS}[A] + 2\pi i w[U] \tag{8}$$

when A_μ changes as

$$A_\mu \to A_\mu^U = U^{-1} A_\mu U + i U^{-1} \partial_\mu U \tag{9}$$

Then, the CS action has to enter with an integer coefficient k if $\exp(k S_{CS}[A])$ is to remain gauge invariant.

As stated before, when fermions couple to the gauge field, a CS term is generated through radiative corrections. The origin of this parity violating contribution (the CS term changes sign under a parity transformation) is twofold. On the one hand, when fermions are massive, there is a parity violation already at the classical level since, in $2+1$ dimensions, a mass term breaks parity. On the other hand, there is a parity anomaly arising in the process of (gauge-invariantly) regularizing the theory. This is why, even for massless fermions, a CS term arises at the quantum level. Different regularization schemes give different answers. This can be seen by computing the fermion determinant using for example a $1/m$ expansion (m being the fermion mass). A Pauli-Villars regularization leads to [8]

$$\log\det(i\partial\!\!\!/ + A\!\!\!/ + m) = \frac{1}{2}\left(\frac{m}{|m|} + \frac{\mu}{|\mu|}\right) S_{CS}[A] + \frac{1}{24\pi|m|}\int d^3x F_{\mu\nu}^2 + O(1/m^2) \tag{10}$$

with μ the Pauli-Villars regulating mass; a zeta function calculation gives instead

$$\log\det(i\partial\!\!\!/ + A\!\!\!/ + m) = \frac{1}{2}\left(\frac{m}{|m|} \pm 1\right) S_{CS}[A] + \frac{1}{24\pi|m|}\int d^3x F_{\mu\nu}^2 + O(1/m^2) \tag{11}$$

the $\pm$ ambiguity in this last formula being related to the way one defines complex powers of the Dirac operators when using Seeley's approach within the zeta function regularization scheme for odd dimensional space-times [31]. We see that an appropriate choice of the fermion mass sign and of the $\pm$ sign (which amounts to a choice of the curve avoiding tha ray of minimal grow of the Dirac operator using zeta function) makes the two answers compatible.

2 Bosonization in $2+1$ dimensions

The method is straightforward. It applies in any number of dimensions but we shall concentrate here in the case of a 3 dimensional manifold M. We shall basically consider the case of free fermions but we shall also discuss an interacting (Thirring) model.

One starts from the fermion Lagrangian for N massive free fermions in

$2+1$ dimensions in the fundamental representation of some group G,

$$L = \bar{\psi}(i\partial\!\!\!/ + m)\psi \tag{12}$$

The corresponding generating functional reads

$$Z_{fer}[s] = \int \mathcal{D}\bar{\psi}\mathcal{D}\psi \exp[-\int d^3x \bar{\psi}(i\partial\!\!\!/ + s\!\!\!/ + m)\psi] \tag{13}$$

where s_μ is the source for fermion currents, $s_\mu = s_\mu^a \tau^a$

Our derivation heavily relies on the invariance of the fermion measure under local gauge transformations $h(x) \in \hat{G}$ with $\hat{G}$ the group of continuous maps $M \to G$. This ensures that

$$Z[s^h] = Z[s] \tag{14}$$

with

$$s_\mu^h = h^{-1} s_\mu h + h^{-1} \partial_\mu h \tag{15}$$

Evidently, fermions can be integated in (13),

$$Z_{fer}[s] = \det(i\partial\!\!\!/ + s\!\!\!/ + m) \tag{16}$$

and the determinant in the r.h.s. of eq.(16) will be used to introduce an auxiliary field b_μ taking values in the Lie algebra of G through the trivial formula

$$Z_{fer}[s] = \int \mathcal{D}b_\mu \, \delta^{(3)}(b_\mu - s_\mu) \, \det(i\partial\!\!\!/ + b\!\!\!/ + m) \tag{17}$$

Now, it will be convenient to replace the delta function in (17) as follows

$$\delta^{(3)}(b_\mu - s_\mu) \to \Delta[b] \, \delta^{(3)} \left(\varepsilon_{\mu_1\mu_2\mu_3}(f_{\mu_1\mu_2}[b] - f_{\mu_1\mu_2}[s])\right) \tag{18}$$

Here we have used that the equation

$$f_{\mu\nu}[b] = f_{\mu\nu}[s] \tag{19}$$

has for $s_\mu \neq 0$ the unique solution

$$b_\mu = s_\mu \tag{20}$$

and $\Delta[b]$ is a Faddeev-Popov-like jacobian,

$$\Delta[b] = |\det(2\varepsilon_{\mu_1\mu_2\mu_3} D_{\mu_1}[b])| \tag{21}$$

with $D_\mu[b]$ the covariant derivative,

$$D_\mu[b] = \partial_\mu + [b_\mu, \ ] \tag{22}$$

(We do not consider for the moment Gribov like problems that could arise for certain manifolds and groups). Were we working in d dimensions an n dimensional ($n = d(d-1)/2$) delta function would be required in the r.h.s. of eq.(18), since one needs for enforcing eq.(19) as many δ-functions as independent components the curvature has.

It is at this point that the bosonic field whose dynamics will be equivalent to that of the original Fermi field comes into play. We introduce it as a Lagrange multiplier A_μ enforcing the δ-function in the path-integral (17),

$$Z_{fer}[s] = \int \mathcal{D}b_\mu \mathcal{D}A_{\mu_3} \det(i\partial\!\!\!/ + b\!\!\!/ + m)\Delta[b]$$
$$\exp\left[\lambda\, tr \int d^dx A_{\mu_3}\varepsilon_{\mu_1\mu_2\mu_3}(f_{\mu_1\mu_2}[b] - f_{\mu_1\mu_2}[s])\right] \tag{23}$$

Here λ is a constant which can be adjusted so as to obtain an adequate normalization for the currents. Now, we rewrite eq.(23) in the form

$$Z_{fer}[s] = \int \mathcal{D}A_{\mu_3} \exp(S_{bos}[A]) \exp\left(-\lambda\, tr \int d^dx A_{\mu_3}\varepsilon_{\mu_1\mu_2\mu_3} f_{\mu_1\mu_2}[s]\right) \tag{24}$$

where the bosonic action is defined as

$$\exp(S_{bos}[A]) = \int \mathcal{D}b_\mu \det(i\partial\!\!\!/ + b\!\!\!/ + m)\Delta[b]$$
$$\exp\left(\lambda\, tr \int d^dx A_{\mu_3}\varepsilon_{\mu_1\mu_2\mu_3} f_{\mu_1\mu_2}[b]\right) \tag{25}$$

Formulae (24)-(25) constitute our basic bosonization recipe: eq.(24) allows to compute fermion current correlation functions in terms of the bosonic field A and eq.(25) gives the bosonic action defining the dynamics of A. It can be now appreciated in what sense we consider our bosonization recipe exact: we have arrived *with no approximation* to a bosonization recipe of the form

$$\bar\psi\gamma_\mu t^a\psi \to 2\,\lambda\,\varepsilon_{\mu\mu_2\mu_3}\partial_{\mu_2}A^a_{\mu_3} \tag{26}$$

Except in $d = 2$ dimensions where we know how to compute exactly the fermion determinant appearing in (25) and to resolve the path-integral defining S_{bos}, one should appeal in the $d = 3$ case (as well as in the case of $d > 3$ dimensions) to some approximation scheme to evaluate the bosonic action accompanying recipe (26). This means that only in $d = 2$ dimensions the complete bosonization recipe is exact.

It should be stressed that the bosonization recipe (26) should be taken as illustrative of the bosonization since the rigorous equivalence between the fermionic and the bosonic theory is at the level of the generating functional

$Z_{fer}[s]$ of Green functions. It is from $Z_{fer}[s]$ written in the form (24) that one has to compute current correlation functions in the bosonic language. Note also that in writing recipe (26) we have ignored terms which are non linear in the source. Although correlation functions of currents acquire a contribution from this terms, this contribution is irrelevant in the calculation of conmutator algebra since they have local support. This can be easily seen, for example, using the BJL method (see [14] for a discussion and [15] for the application of the BJL method within the present bosonization approach).

Exactly the same steps lead in d dimensions to a formula analogous to (26) which reads

$$\bar{\psi}\gamma_\mu t^a \psi \to 2\,\lambda\,\varepsilon_{\mu\mu_2\ldots\mu_d}\partial_{\mu_2}A^a_{\mu_3\ldots\mu_d} \tag{27}$$

Here the bosonic field $A^a_{\mu_3\ldots\mu_d}$ corresponds to scalar fields in $d=2$ dimensions (see ref.[14] for details on how to make contact with the usual bosonization rules), to the vector field already introduced in $d=3$ dimensions and to an antisymmetric (Kalb-Rammond) field in $d>3$ dimensions [11,13].

3 The abelian case in $d=3$

The Abelian case in 3 dimensions is particularly simple. To begin with, the bosonization recipe for the fermion current reads

$$\bar{\psi}\gamma_\mu\psi \to \pm\frac{i}{8\pi}\varepsilon_{\mu\nu\alpha}\partial_\nu A_\alpha \tag{28}$$

where we have chosen λ so as to make contact with the normalization employed in ref.[15]. Concerning the bosonic action, $\Delta[b]$ is trivial so that eq.(25) simply reads

$$\exp(S_{bos}[A]) = \int \mathcal{D}b_\mu \det(i\partial\!\!\!/ + b\!\!\!/ + m)\exp\left(\mp\frac{i}{16\pi}\,tr\int d^3x A_\mu\varepsilon_{\mu\nu\alpha}f_{\nu\alpha}[b]\right) \tag{29}$$

or, calling

$$-\log\det(i\partial\!\!\!/ + b\!\!\!/ + m) = \int d^3x L[b] \tag{30}$$

we can write

$$\exp(S_{bos}[A]) = \int \mathcal{D}b_\mu \exp(-S_{eff}[b,A]) \tag{31}$$

where S_{eff} is defined as

$$S_{eff}[b,A] = \int d^3x(L[b] \pm \frac{i}{16\pi}\,tr\,A_\mu\varepsilon_{\mu\nu\alpha}f_{\nu\alpha}[b]) \tag{32}$$

(The double sign in eqs.(29)-(32) is included for convenience, see the discussion below)

Being in general $L[b]$ non-quadratic in b one cannot path-integrate in (29) so as to obtain $S_{bos}[A]$. We shall see however that there is a change of variables allowing to decouple A_μ from b_μ in $S_{eff}[b, A]$ so that one can control the A_μ dependence of $S_{bos}[A]$ without necessity of explicitly integrating over b_μ. Let us define a new variable b'_μ through the equation

$$b_\mu = (1-\theta)b'_\mu + \theta A_\mu + V_\mu[A] \tag{33}$$

where $V_\mu[A]$ is a gauge invariant function of A_μ so that b'_μ, the variable which will replace b_μ, transforms as a a gauge field. θ is an arbitrary parameter to be adjusted later. The idea is to choose V_μ so as to decouple b'_μ from A. This amounts to impose the following condition

$$\frac{\delta^2 S_{eff}}{\delta b'_\mu(x)\delta A_\nu(y)} = 0 \tag{34}$$

which in terms of V_μ reads

$$2i\lambda\varepsilon_{\rho\sigma\alpha}\partial^\alpha_x\delta(x-y) + \theta_y \int d^3z \frac{\delta^2 L[b(z)]}{\delta b_\rho(y)\delta b_\sigma(x)}$$
$$+ \int d^3u \left(\int d^3z \frac{\delta^2 L[b(z)]}{\delta b_\beta(u)\delta b_\sigma(x)}\right) \frac{\delta V_\beta(u)}{\delta A_\rho(y)} = 0 \tag{35}$$

To go on we need an explicit and necessarily approximate expression for $L[b]$. If we are interested in the large-distance regime of the bosonic theory we can use the result first [5]-[8] for the $d = 3$ fermion determinant as an expansion in inverse powers of the fermion mass

$$L[b] = \mp\frac{i}{8\pi}\varepsilon_{\mu\nu\alpha}b_\mu\partial_\nu b_\alpha + \frac{1}{24\pi|m|}f^2_{\mu\nu}[b] + O(\frac{1}{m^2}) \tag{36}$$

The first term in (36) is the well-honored Chern-Simons action introduced in [5] as a way of generating a mass for gauge fields in three dimensions. The double sign in this term is originated in a regularisation ambiguity characteristic of odd-dimensions (see ref.[31]). The second one corresponds to the leading parity-even contribution to the fermion determinant. One easily sees that if one tries for $V_\mu[A]$ the functional form

$$V_\mu[A] = i\frac{C}{m}\varepsilon_{\mu\nu\alpha}f_{\nu\alpha}[A] \tag{37}$$

one gets, from the decoupling equation (35),

$$C = \pm 1/3 \tag{38}$$

Then, if for simplicity one chooses $\theta = -1$, the bosonic action for A_μ can be easily found to be

$$S_{bos}[A] = \pm\frac{i}{8\pi}\int d^3x\, \varepsilon_{\mu\nu\alpha}A_\mu\partial_\nu A_\alpha + \frac{1}{24\pi|m|}\int d^3x\, f^2_{\mu\nu}[A] + O(1/m^2). \quad (39)$$

One can in principle determine, following the same procedure, the following terms in the $1/m$ expansion of S_{bos} by including the corresponding terms in the fermion determinant expansion. This result extends that originally presented in ref.[9]. It shows that the bosonic counterpart of the three dimensional free fermionic theory is, to order $1/m^2$, a Maxwell-Chern-Simons theory which is equivalent, in turn, to a self-dual system [32-33].

Alternatively to the $1/m$ determinant expansion, one can consider an expansion in powers of b_μ retaining up to quadratic terms. The result can be written in the form [28]

$$L[b] = \frac{i}{2}\varepsilon_{\mu\nu\alpha}b_\mu P\partial_\nu b_\alpha + \frac{1}{4|m|}f_{\mu\nu}[b]Qf_{\mu\nu}[b] \quad (40)$$

where P and Q are functionals to be calculated within a loop expansion,

$$P \equiv P\left(\frac{\partial^2}{m^2}\right) \qquad Q \equiv Q\left(\frac{\partial^2}{m^2}\right) \quad (41)$$

Details of the calculations of P and Q and results within the loop-expansion can be found in refs. [28,34].

In order to decouple the b_μ field one again proposes a change of variables like in (33) but now trying for V_μ the (gauge-invariant) functional form

$$V_\mu[A] = \frac{i}{m}\varepsilon_{\mu\nu\alpha}Rf_{\nu\alpha}[A] = 2\frac{i}{m}\varepsilon_{\mu\nu\alpha}R\partial_\nu A_\alpha \quad (42)$$

with

$$R \equiv R\left(\frac{\partial^2}{m^2}\right) \quad (43)$$

One finds, from the decoupling conditions (35),

$$\frac{\delta^2 S_{eff}[b', A]}{\delta A_\rho(y)\delta b'_\sigma(x)} =$$

$$(1-\theta)\left(i\varepsilon_{\rho\sigma\alpha}\left(2\lambda + \theta P - 2\frac{\partial^2}{m^2}QR\right)\partial_\alpha\delta(x-y)\right.$$

$$\left.+\frac{2}{m}\left(\frac{1}{2}\theta Q - PR\right)\left(\partial_\rho\partial_\sigma - \delta_{\rho\sigma}\partial^2\right)\delta(x-y)\right) = 0 \quad (44)$$

θ being here a functional of ∂^2/m^2. The solution of this equation is

$$R = -\lambda \frac{Q}{(P^2 - \frac{\partial^2}{m^2}Q^2)} \qquad \theta = -2\lambda \frac{P}{(P^2 - \frac{\partial^2}{m^2}Q^2)} \tag{45}$$

With this choice, the change of variables decouples the b_μ integration so that one can finally get the bosonic action for A_μ which now reads

$$S_{bos}[A] = \int d^3x \left(-(2\lambda)^2 \frac{i}{2} \varepsilon_{\mu\nu\alpha} A_\mu \frac{P}{(P^2 - \frac{\partial^2}{m^2}Q^2)} \partial_\nu A_\alpha \right.$$
$$\left. +(2\lambda)^2 \frac{1}{4m} f_{\mu\nu}[A] \frac{Q}{(P^2 - \frac{\partial^2}{m^2}Q^2)} f_{\mu\nu}[A] \right) \tag{46}$$

This result coincides with that found in ref.[28], obtained by a direct functional integration on b_μ. As it was proven in this last work, it corresponds for massless fermions to the bosonization action proposed in ref.[26] since in the $m \to 0$ limit eq.(46) takes the form

$$S_{bos} = \frac{2}{\pi} \int d^3x \left(\frac{1}{4} f_{\mu\nu} \frac{1}{\sqrt{-\partial^2}} f_{\mu\nu} - \frac{i}{2} \epsilon_{\mu\nu\lambda} A_\mu \partial_\nu A_\lambda \right) \tag{47}$$

4 Interacting models

One can apply the bosonization approach described above to analyse interacting fermionic models. Let us consider for example the Thirring model with a current-current interaction Lagrangian of the form

$$L_{int} = -\frac{g^2}{2N} j_\mu j_\mu \tag{48}$$

where ψ^i are N two-component Dirac spinors and j^μ the $U(1)$ current,

$$j_\mu = \bar{\psi}^i \gamma^\mu \psi^i. \tag{49}$$

The coupling constant g^2 has dimensions of inverse mass. (Although non-renormalizable by power counting, four fermion interaction models in $2+1$ dimensions are known to be renormalizable in the $1/N$ expansion [35].) One can directly apply the bosonization recipe found in the precedent section to this interaction Lagrangian. Making a choice of λ so as to coincide with the normalization in [9], this meaning

$$j_\mu \to i\sqrt{\frac{N}{4\pi}} \varepsilon_{\mu\nu\alpha} \partial_\nu A_\alpha \tag{50}$$

one has

$$L_{int} \to \frac{g^2}{16\pi} f_{\mu\nu}^2 \tag{51}$$

Alternatively, one can eliminate the quartic fermionic interaction by introducing an auxiliary field a_μ via the identity

$$\exp\Big(\int \frac{g^2}{2N} j^\mu j_\mu d^3x\Big) = \int \mathcal{D}a_\mu \exp\Big[-\int \Big(\frac{1}{2}a^\mu a_\mu + \frac{g}{\sqrt{N}} j^\mu a_\mu\Big) d^3x\Big] \tag{52}$$

and then proceed to integrate fermions as in the free case thus obtaining a determinant in which the a_μ field can be eliminated by a shift $b_\mu \to b_\mu - a_\mu$. One confirms in this way that the bosonization recipe (51) is correct so that the three dimensional Thirring model is equivalent, in the $1/m$ approximation to a Maxwell-Chern-Simons model. To leading order in $1/m$ we can then write (after rescaling the field A_μ)

$$L_{Th} \to \frac{1}{4} f_{\mu\nu}^2 \pm i \frac{2\pi}{g^2} \epsilon^{\mu\alpha\nu} A_\mu \partial_\alpha A_\nu \tag{53}$$

We can give now a first application of the bosonization formulas and, in this way, explore their physical content. The Lagrangian in (53) has a Chern-Simons term which controls its long distance behavior. It is well known[22,7] that the Chern-Simons gauge theory is a theory of knot invariants which realizes the representations of the Braid group. These knot invariants are given by expectation values of Wilson loops in the Chern-Simons gauge theory. In this way, it is found that the expectation values of the Wilson loop operators imply the existence of excitations with fractional statistics. Thus, it is natural to seek the fermionic analogue of the Wilson loop operator W_Γ which, in the Maxwell-Chern-Simons theory is given by

$$W_\Gamma = \langle \exp\{i \frac{\sqrt{N}}{g} \oint_\Gamma A_\mu dx^\mu\}\rangle \tag{54}$$

where Γ is the union of a an arbitrary set of closed curves (loops) in three dimensional euclidean space. Given a closed loop (or union of closed loops) Γ, it is always possible to define a set of open surfaces Σ whose boundary is Γ, *i.e.* $\Gamma = \partial\Sigma$. Stokes' theorem implies

$$\begin{aligned} \langle \exp\{i \frac{\sqrt{N}}{g} \oint_\Gamma A_\mu dx^\mu\}\rangle &= \langle \exp\{i \frac{\sqrt{N}}{g} \int_\Sigma dS_\mu \epsilon^{\mu\nu\lambda} \partial_\nu A_\lambda\}\rangle \\ &= \langle \exp\{i \frac{\sqrt{N}}{g} \int d^3x\, \epsilon^{\mu\nu\lambda} \partial_\nu A_\lambda\, b_\lambda\}\rangle \end{aligned} \tag{55}$$

Here $b_\lambda(x)$ is the vector field

$$b_\lambda(x) = n_\lambda(x)\delta_\Sigma(x) \tag{56}$$

with n_λ a field of unit vectors normal to the surface Σ and $\delta_\Sigma(x)$ is a delta function with support on Σ. Using eq.(50) we find that this expectation value becomes, in the Thirring Model, equivalent to

$$W_\Gamma = \langle \exp\{i\frac{\sqrt{N}}{g}\oint_{\partial\Sigma} dx_\mu A^\mu\}\rangle_{MCS} = \langle \exp\{\int_\Sigma dS_\mu \bar{\psi}\gamma^\mu\psi\}\rangle_{Th} \tag{57}$$

More generally we find that the Thirring operator $\mathcal{W}_\Sigma$

$$\mathcal{W}_\Sigma = \langle \exp\{q\int_\Sigma dS_\mu \bar{\psi}\gamma^\mu\psi\}\rangle_{Th} \tag{58}$$

obeys the identity

$$\langle \exp\{q\int_\Sigma dS_\mu \bar{\psi}\gamma^\mu\psi\}\rangle_{Th} = \langle \exp\{iq\frac{\sqrt{N}}{g}\oint_{\partial\Sigma} A_\mu dx^\mu\}\rangle_{MCS} \tag{59}$$

for an arbitrary fermionic charge q.

The identity (59) relates the flux of the fermionic Thirring current through an open surface Σ with the Wilson loop operator associated with the boundary Γ of the surface. The Wilson loop operator can be trivially calculated in the Maxwell-Chern-Simons theory. For very large and smooth loops the behavior of the Wilson loop operators is dominated by the Chern-Simons term of the action. The result is a topological invariant which depends only on the linking number ν_Γ of the set of curves Γ [22,7]. By an explict calculation one finds

$$\langle \exp\{q\int_\Sigma dS_\mu \bar{\psi}\gamma^\mu\psi\}\rangle_{Th} = \exp\{\mp i\nu_\Gamma \frac{Nq^2}{8\pi}\} \tag{60}$$

This result implies that the non-local Thirring loop operator $\mathcal{W}_\Sigma$ exhibits fractional statistics with a statistical angle $\delta = Nq^2/8\pi$. The topological significance of this result bears close resemblance with the bosonization identity in $1+1$ dimensions between the circulation of the fermionic current on a closed curve and the topological charge (or instanton number) enclosed in the interior of the curve [36]. From the point of view of the Thirring model, this is a most surprising result whthe power of the bosonization identities. To the best of our knowledge, this is the first example of a purely fermionic operator, albeit non-local, which is directly related to a topological invariant.

5 Current algebra

We have seen in precedent sections that in 3 dimensional space-time the fermion action bosonizes, in the large m limit, to a Maxwell-Chern-Simons theory. Now, the gauge invariant algebra of such theory has been studied in refs.[5,33]. One has for instance (with our conventions),

$$[E_i(\vec{x},t), B(\vec{y},t)] = -3\,|m|\epsilon_{ij}\partial_j\delta^{(2)}(\vec{x}-\vec{y}) \tag{61}$$

If one now relates the electric field $E_i = F_{i0}$ and the magnetic field $B = \epsilon_{ij}\partial_i A_j$ to the fermionic currents through the bosonization recipe for the fermion current,

$$j_o \to \frac{1}{\sqrt{4\pi}} B \tag{62}$$

$$j_i \to \frac{1}{\sqrt{4\pi}} \epsilon_{ij} E_j \tag{63}$$

then, the resulting fermion current commutator algebra is not the one to be expected for three-dimensional free fermions. Indeed, the $d = 3$ fermion current algebra should contain an infinite Schwinger term [40-41] which is absent in eq.(61). The point is that calculations leading to a bosonic theory of the Maxwell-Chern-Simons type are valid only for large fermion mass while calculation of equal-time current commutators imply, as we shall see, a limiting procedure which cannot be naively followed for large masses.

Since the exact bosonic partition function is much too complicated to handle, a possible strategy is to use the quadratic (in auxiliary fields) approximation mentioned in the precedent section working with an arbitrary (not necessarily large) mass so as to obtain a bosonized version of the original fermionic model in which the equal-timOne should then compute current commutators for this bosonized theory, and test whether they coincide with those satisfied by fermionic currents in the original model. Details of this calculation can be found in [12]. I will just sketch here the principal steps leading to the consistent equal-time current commutators in the bosonic language.

As explained above, within the quadratic (in b_μ) approximation, one can write the fermionic partition function in terms of the bosonic fields A_μ in the form [28]

$$Z_{fer} = \int DA_\mu \exp\left[-\int d^3x(\frac{1}{4}F_{\mu\nu}C_1F_{\mu\nu} - \frac{i}{2}A_\mu C_2\epsilon_{\mu\nu\lambda}\partial_\nu A_\lambda \right.$$
$$\left. +is_\mu\epsilon_{\mu\nu\lambda}\partial_\nu A_\lambda)\right] \tag{64}$$

with C_1 and C_2 now given through their momentum-space representation $\tilde{C}_1$ and $\tilde{C}_2$

$$\tilde{C}_1(k) = \frac{1}{4\pi}\frac{\tilde{F}(k)}{k^2\tilde{F}^2(k)+\tilde{G}^2(k)} \tag{65}$$

$$\tilde{C}_2(k) = \frac{1}{4\pi}\frac{\tilde{G}(k)}{k^2\tilde{F}^2(k)+\tilde{G}^2(k)} \tag{66}$$

and $\tilde{F}(k)$ and $\tilde{G}(k)$ given by [28]

$$\tilde{F}(k) = \frac{|m|}{4\pi k^2}\left[1 - \frac{1-\frac{k^2}{4m^2}}{(\frac{k^2}{4m^2})^{\frac{1}{2}}}\arcsin(1+\frac{4m^2}{k^2})^{-\frac{1}{2}}\right],$$

$$\tilde{G}(k) = \frac{q}{4\pi} + \frac{m}{2\pi |k|}\arcsin(1+\frac{4m^2}{k^2})^{-\frac{1}{2}}$$

Being quadratic in A_μ, eq.(64) can be integrated leading to

$$Z_{bos}[s] = [detD_{\mu\nu}]^{-\frac{1}{2}}\exp\left[\frac{1}{8\pi}\int d^3x d^3y \partial_\nu s_\mu(x)\epsilon_{\mu\nu\lambda}D^{-1}_{\lambda\rho}(x,y)\partial_\sigma s_\tau(y)\epsilon_{\rho\sigma\tau}\right] \tag{67}$$

where $D^{-1}_{\mu\nu}$ is just the propagator of the bosonic action which, in the Lorentz gauge we adopt from here on, reads

$$D^{-1}_{\mu\nu}(x,y) = \int\frac{d^3k}{(2\pi)^3}\left[P(k)g_{\mu\nu} + Q(k)k_\mu k_\nu + R(k)\epsilon_{\mu\nu\alpha}k_\alpha\right]\exp ik(x-y) \tag{68}$$

with

$$P(k) = \frac{\tilde{C}_1(k)}{k^2\tilde{C}_1^2(k)+\tilde{C}_2^2(k)} = 4\pi\tilde{F}(k) \tag{69}$$

$$Q(k) = \frac{\tilde{C}_1(k)}{k^2\tilde{C}_1^2(k)+\tilde{C}_2^2(k)}\left(\frac{\tilde{C}_2(k)}{k^2\tilde{C}_1(k)}\right)^2 \tag{70}$$

$$R(k) = \frac{\tilde{C}_2(k)}{k^2(k^2\tilde{C}_1^2(k)+\tilde{C}_2^2(k))} \tag{71}$$

Let us briefly recall how one can compute current commutators within the path-integral scheme using the so-called BJL method [37-39]. To this end we define the correlator

$$G_{\mu\nu}(x,y) = \left.\frac{\delta^2 log Z_{fer}[s]}{\delta s_\mu(x)\delta s_\nu(y)}\right|_{s=0} \tag{72}$$

from which one can easily derive equal time current commutators using the relation

$$< [j_0(\vec{x},t), j_i(\vec{y},t)] >= \lim_{\epsilon\to 0+} [G_{0i}(\vec{x},t+\epsilon;\vec{y},t) - G_{0i}(\vec{x},t-\epsilon;\vec{y},t)] \tag{73}$$

The current commutator evaluated using eqs.(72)-(73) corresponds to $Z_{fer}[s]$ written in terms of bosonic fields. That is, eq.(73) gives the equal-time commutator for the bosonic currents $j_\mu = (1/\sqrt{4\pi})\epsilon_{\mu\nu\alpha}\partial_\nu A_\alpha$. This result should then be compared with that arising in the original 3-dimensional fermionic model for which $j_\mu = -i\bar{\psi}\gamma_\mu\psi$ [41].

Starting from eqs.(67)-(68) and using the BJL method we get, after some calculations,

$$G_{\mu\nu}(x,y) = -\frac{1}{4\pi}\epsilon_{\mu\alpha\rho}\epsilon_{\nu\beta\sigma}\partial_\alpha\partial_\beta D^{-1}_{\rho\sigma} \tag{74}$$

or

$$G_{\mu\nu}(x,y) = \frac{1}{4\pi}\int \frac{d^3k}{(2\pi)^3}[P(k)(k^2 g_{\mu\nu} - k_\mu k_\nu) + k^2 R(k)\epsilon_{\mu\nu\alpha}k_\alpha] \exp[ik(x-y)] \tag{75}$$

With this, we can rewrite eq.(73) in the form

$$< [j_0(\vec{x},t), j_i(\vec{y},t] >= \lim_{\epsilon\to 0+} I^\epsilon(\vec{x}-\vec{y}) \tag{76}$$

with

$$I^\epsilon(\vec{x}) = -2i\int \frac{d^3k}{(2\pi)^3} k_0 k_i \sin(k_0\epsilon)\tilde{F}(k)\exp(i\vec{k}.\vec{x}) \tag{77}$$

where we have written $(k_\mu) = (k_o, k_i)$, $i = 1,2$. It will be convenient to define

$$k_0' = \epsilon k_0 \tag{78}$$

In terms of this new variable and using the explicit form for $\tilde{F}(k)$ given in [28,12], with $k = (k_0^2 + \vec{k}^2)^{1/2}$, integral I^ϵ becomes

$$I^\epsilon(\vec{x}) = -\frac{1}{8\pi^2|m|}\frac{1}{\epsilon^2}\partial_i \int \frac{d^2k}{(2\pi)^2}\exp i(\vec{k}.\vec{x})\int_0^\infty dk_0' k_0' \sin k_0' f(y) \tag{79}$$

where

$$f(y) = \frac{1}{y}\left[1 - \frac{(1-y)}{\sqrt{y}}\arcsin\frac{1}{\sqrt{1+(1/y)}}\right] \tag{80}$$

and we have defined

$$y = \frac{k^2}{4m^2} = \frac{k'^2_0 + \epsilon^2\vec{k}^2}{4\epsilon^2 m^2} \tag{81}$$

One can now see that $y \to \infty$ for $\epsilon \to 0$ and fixed m. Then, expanding in powers of $1/y$ one has $f(y) \sim \pi/(2\sqrt{y})$ and then using distribution theory to define the integral over k'_0 one finds

$$< [j_0(\vec{x},t), j_i(\vec{y},t] >= -\frac{1}{8\pi}\lim_{\epsilon\to 0}\frac{1}{\epsilon}\partial_i\delta^{(2)}(\vec{x}-\vec{y}) \tag{82}$$

This result for the equal-time current commutator, evaluated within the bosonized theory, shows exactly the infinite Schwinger term that is found, using the BJL method, for free fermions in $d = 3$ dimensions [41]. As it happens in $d = 4$ dimensions [42], we see from eq.(82) that the commutator at *unequal* times is well defined: divergencies appear only when one takes the equal-time limit.

One can evaluate also the next order vanishing in the equal-time current commutator so as to compare it with the result from the original fermion model reported in the literature [41]. The answer is [12]

$$< [j_0(\vec{x},t), j_i(\vec{y},t)] > = -\frac{1}{8\pi}\lim_{\epsilon\to 0}\left(\frac{1}{\epsilon}\partial_i\delta^{(2)}(\vec{x}-\vec{y})\right.$$
$$\left. -\frac{\epsilon}{\Lambda}[4m^2\partial_i\delta^{(2)}(\vec{x}-\vec{y}) - \frac{1}{2}\partial_i\Delta\delta^{(2)}(\vec{x}-\vec{y})]\right) \tag{83}$$

where we have defined

$$\frac{1}{\Lambda} = \int_0^\infty dk'_0\frac{1}{k'^2_0}\sin k'_0 \tag{84}$$

In order to compare with ref.[41] where current commutators were computed using dimensional regularization, we define, coming back to the original variable $k_0 = k'_0/\epsilon$

$$A[d] = \frac{1}{2}\int d^{d-2}k_0\frac{1}{k_0^2} sin k_0\epsilon \tag{85}$$

so that $A[d=3] = \epsilon/\Lambda$. One can now perform the analytically continued integral to find, near $d = 3$, the behavior

$$A[d] \sim -\epsilon \times \frac{\epsilon^{3-d}}{3-d} \tag{86}$$

The same ambiguous result for free fermions is obtained in ref.[41] near $d = 3$. This ambiguity can be however removed, the pole in dimensional regularization corresponding as usual to a logarithmic divergence. It is also interesting to note that if one uses the nice approximation $\tilde{F}_{appr}$ for $\tilde{F}$ proposed in ref.[28], one can well check the correctness of our previous analysis [12].

¿From the analysis above, we see that not only the infinite Schwinger term, analogous to that arising in $d = 4$ [42] is obtained in the bosonized version of our $d = 3$ fermion theory but also the mass-dependent second term as well as the triple derivative third term, both vanishing in the equal time limit. One can see that in the large mass regime, terms depending on the product $\epsilon m = \lambda$ will produce ambiguities according to the way both limits ($\epsilon \to 0$ and $m \to \infty$) are taken into account, a problem which is not present in the limit of small masses. To see this in more detail, let us come back to (77) and consider the case in which λ is kept fixed while $\epsilon \to 0$ (so that $m \to \infty$). In this case, taking the limit before integrating out k_0', one finds for I^ϵ

$$I^\epsilon(\vec{x}) \sim |m| h(\lambda) \partial_i \delta^{(2)}(\vec{x}) \cdot \tag{87}$$

where

$$h(\lambda) = \frac{1}{2\pi} \int_0^\infty dz z \sin(2\lambda z) f(z) \tag{88}$$

with f given by eq.(80). Let us note that using the approximate $\tilde{F}_{appr}$ of ref.[28] and taking the limit after the exact integration over k_0', we recover the same behavior (87). We see that for $\lambda = \epsilon m$ fixed, h just gives a normalization factor so that one reproduces from I^ϵ in the form (87) a commutator algebra at equal times and large mass that coincides with that to be infered from a Maxwell-Chern-Simons theory,

$$< [j_0(\vec{x}, t), j_i(\vec{y}, t] > \longrightarrow c|m| \partial_i \delta^{(2)}(\vec{x} - \vec{y}) \quad (m \to \infty) \tag{89}$$

with c a normalization constant. Again, currents appearing in the l.h.s. of eq.(89) are bosonic currents which can be written in terms of the electric and magnetic fields thus reproducing the MCS gauge invariant algebra [5,33]. One should note however that the free fermion - MCS mapping is valid in the large mass limit of the original fermionic theory, this meaning the large-distances regime for fermion fields. Since current commutators test the short-distance regime, one should not take the MCS gauge-invariant algebra as a starting point to reproduce the fermion current commutators.

6 The non-Abelian case in $d = 3$

In three dimensional space-time, the bosonic action (25) takes the form

$$\exp(-S_{bos}[A]) = \int \mathcal{D}b_\mu \mathcal{D}\bar{c}_\mu \mathcal{D}c_\mu \exp\left(-tr\int d^3x \, (L[b] \pm \frac{i}{8\pi}\varepsilon_{\mu\nu\alpha}\bar{c}_\mu D_\nu[b]c_\alpha \right.$$
$$\left. \mp \frac{i}{16\pi}(A_\mu - b_\mu)^* f_\mu[b])\right) \tag{90}$$

Here ghost fields $\bar{c}_\mu$ and c_μ were introduced to represent the Faddeev-Popov like determinant $\Delta[b]$. Again, we have written

$$tr\int d^3x \, L[b] = -\log\det(i\partial\!\!\!/ + m + b\!\!\!/) \tag{91}$$

and we have chosen the arbitrary constant λ appearing in (25) so as to make contact with the conventions of ref.[15], $\lambda = \frac{i}{16\pi}$. Moreover, we have shifted the bosonic field $A_\mu \to A_\mu - b_\mu$ (this amounting to a trivial Jacobian) for reasons that will become clear below.

It was observed in ref.[14] that when $L[b]$ is approximated by its first term in the $1/m$ expansion, a set of BRST transformations can be defined so that the corresponding BRST invariance allows to obtain the (approximate) bosonic action. We shall explicitly prove here that this invariance is present in (90) where no approximation for $L[b]$ is assumed. To this end, we introduce a set of auxiliary fields h_μ (taking values in the Lie algebra of G), l and $\bar{\chi}$ so that one can rewrite (90) in the form

$$\exp(-S_{bos}[A]) = \int \mathcal{D}b_\mu \mathcal{D}\bar{c}_\mu \mathcal{D}c_\mu \mathcal{D}h_\mu \mathcal{D}l \mathcal{D}\bar{\chi} \exp(-S_{eff}[A, b, h, l, \bar{c}, c, \bar{\chi}]) \tag{92}$$

with

$$S_{eff}[A, b, h, l, \bar{c}, c, \bar{\chi}] = tr\int d^3x \, (L[b - h] \pm \frac{i}{8\pi}\varepsilon_{\mu\nu\alpha}\bar{c}_\mu D_\nu[b]c_\alpha$$
$$\mp \frac{i}{16\pi}((A_\mu - b_\mu)^* f_\mu[b] + lh_\mu^2 - 2\bar{\chi}h_\mu c_\mu)) \tag{93}$$

where $\bar{\chi}$ is an anti-ghost field. Written in the form (93), the bosonic action has a BRST invariance under the following nilpotent off-shell BRST transformations

$$\delta\bar{c}_\mu = A_\mu - b_\mu, \qquad \delta b_\mu = c_\mu, \qquad \delta A_\mu = c_\mu, \qquad \delta c_\mu = 0, \qquad \delta\bar{\chi} = l$$

$$\delta h_\mu = c_\mu, \qquad \delta l = 0 \tag{94}$$

In view of this BRST invariance, one could add to S_{eff} a BRST exact form without changing the dynamics defined by $S_{bos}[A]$. Exploiting this, we shall see that one can factor out the A_μ dependence in the r.h.s. of eq.(93) so that it completely decouples from the path-integral over b_μ auxiliary and ghost fields exactly as we did in the Abelian case. Although complicated, this integral then becomes irrelevant for the definition of the bosonic action for A_μ. Indeed, let us add to S_{eff} the BRST exact form δG,

$$S_{eff}[A,b,h,l,\bar{c},c,\chi] \to S_{eff}[A,b,h,l,\bar{c},c,\chi] + \delta G[A,b,h,\bar{c}] \tag{95}$$

with

$$G[A,b,h,\bar{c}] = \mp\frac{i}{16\pi} tr \int d^3x\, \varepsilon_{\mu\nu\alpha}\bar{c}_\mu H_{\nu\alpha}[A,b,h] \tag{96}$$

and $H_{\nu\alpha}[A,b,h]$ a functional to be determined in order to produce the decoupling. Then, consider the change of variables (analogous to (33) for the Abelian case)

$$b_\mu = 2b'_\mu - A_\mu + V_\mu[A] \tag{97}$$

where $V_\mu[A]$ is some functional of A_μ changing covariantly under gauge transformations,

$$V_\mu[A^g] = g^{-1}V_\mu[A]g \tag{98}$$

so that b'_μ is, like A_μ and b_μ, a gauge field. Integrating over l in (92) and imposing the resulting constraint, $h_\mu = 0$, one sees that if one imposes on $H_{\nu\alpha}[A,b,h]$ the condition

$$\varepsilon_{\mu\nu\alpha}\int d^3y \left(\frac{\delta H_{\nu\alpha}}{\delta b^a_\rho(y)} + \frac{\delta H_{\nu\alpha}}{\delta A^a_\rho(y)} + \frac{\delta H_{\nu\alpha}}{\delta h^a_\rho(y)}\right) c^a_\rho(y)\,|_{h=0} = \varepsilon_{\mu\nu\rho}[A_\nu - b_\nu - V_\nu[A], c_\rho] \tag{99}$$

then, when written in terms of the new b'_μ variable, the ghost term becomes

$$S_{ghost}[b',c,\bar{c}] = \pm\frac{i}{8\pi} tr \int d^3x\, \varepsilon_{\mu\nu\alpha}\bar{c}_\mu D_\nu[b']c_\alpha \tag{100}$$

so that its contribution is still A_μ independent. Then, we can write the effective action in the form

$$S_{eff}[b',A] + S_{ghost}[b',c,\bar{c}] \tag{101}$$

with

$$\begin{aligned} S_{eff}[b',A] &= \tilde{S}[b,A] \\ &= tr\int d^3x \left(L[b] \mp \frac{i}{16\pi}(A_\mu - b_\mu)({}^*f_\mu[b] + {}^*H_\mu[A,b,0])\right) \end{aligned} \tag{102}$$

where ${}^*H_\mu = \varepsilon_{\mu\nu\alpha} H_{\nu\alpha}$.

Condition (99) made the ghost term independent of the bosonic field A_μ. We shall now impose a second constraint in order to completely decouple the auxiliary field b'_μ from A_μ in S_{eff}. Indeed, consider the conditions

$$\frac{\delta^2 S_{eff}[b', A]}{\delta A^a_\rho(y) \delta\, b'^b_\sigma(x)} = 0 \tag{103}$$

In terms of the original auxiliary field b_μ these equations read

$$\frac{\delta^2 \tilde{S}[b, A]}{\delta A^a_\rho(y) \delta b^b_\sigma(x)} - \frac{\delta^2 \tilde{S}[b, A]}{\delta b^a_\rho(y) \delta b^b_\sigma(x)} + \int d^3u \frac{\delta^2 \tilde{S}[b, A]}{\delta b^c_\beta(u) \delta b^b_\sigma(x)} \frac{\delta V^c_\beta(u)}{\delta A^a_\rho(y)} = 0 \tag{104}$$

Eqs.(104) can be easily written in terms of L, H and V as a lengthy equation that we shall omit here.

The strategy is now as follows: once a given approximate expression for the fermion determinant is considered, one should solve eq.(104) in order to determine functionals V in eq.(97) and G in eq.(96), taking also in account the condition (99). In particular, if one considers the $1/m$ expansion for the fermion determinant, equations (99) and (104) should determine the form of V and G as a power expansion in $1/m$. In ref.[8] the $1/m$ expansion for the fermion determinant was shown to give

$$\ln \det(i\partial\!\!\!/ + m + b\!\!\!/) = \pm \frac{i}{16\pi} S_{CS}[b] + I_{PC}[b] + O(\partial^2/m^2), \tag{105}$$

where the Chern-Simons action S_{CS} is given by

$$S_{CS}[b] = \varepsilon_{\mu\nu\lambda}\, \mathrm{tr} \int d^3x\, (f_{\mu\nu} b_\lambda - \frac{2}{3} b_\mu b_\nu b_\lambda). \tag{106}$$

Concerning the parity conserving contributions, one has

$$I_{PC}[b] = -\frac{1}{24\pi m}\, \mathrm{tr} \int d^3x\, f^{\mu\nu} f_{\mu\nu} + \cdots, \tag{107}$$

To order zero in this expansion, solution of eqs.(99),(104) is very simple. Indeed, in this case the fermion determinant coincides with the CS action and one can easily see that the solution is given by

$$V^{(0)}_\mu[A] = 0 \tag{108}$$

$$G^{(0)}[A, b, h, \bar{c}] = \pm \frac{i}{16\pi} tr \int d^3x\, \bar{c}_\mu\, (\frac{1}{2}\, {}^*f_\mu[A] + \frac{1}{2}\, {}^*f_\mu[b] - 2\, {}^*D_{\mu\alpha}[A] h_\alpha) \tag{109}$$

With this, the change of variables (97) takes the simple form

$$b_\mu = 2b'_\mu - A_\mu \tag{110}$$

and the decoupled effective action reads

$$S^{(0)}_{eff}[b,A,\bar{c},c] = \mp\frac{i}{16\pi}(2S_{CS}[b'] - S_{CS}[A]) + S_{ghost}[b'] \tag{111}$$

We then see that the path-integral defining the bosonic action $S_{bos}[A]$, factors out so that one ends with a bosonic action in the form

$$S^{(0)}_{bos}[A] = \pm\frac{i}{16\pi}S_{CS}[A] \tag{112}$$

as advanced in [10,14]. Let us remark that in finding the solution for G one starts by writing the most general form compatible with its dimensions,

$$G^{(0)}[A,b,h,\bar{c}] = tr\int d^3x\, \varepsilon_{\mu\nu\alpha}\bar{c}_\mu\,(d_1 b_\nu A_\alpha + d_2 A_\nu b_\alpha + d_3 b_\nu b_\alpha + d_4 A_\nu A_\alpha$$
$$+d_5 b_\nu h_\alpha + d_6 h_\nu b_\alpha + d_7 A_\nu h_\alpha + d_8 h_\nu A_\alpha + d_9\partial_\nu A_\alpha + d_{10}\partial_\nu b_\alpha + d_{11}\partial_\nu h_\alpha) \tag{113}$$

All the arbitrary parameters d_i are determined by imposing the conditions (99) and (104) with ${}^*H_\mu$ transforming covleads, together with a gauge invariant action, to the solution (109).

To go further in the $1/m$ expansion one uses the next to the leading order in the fermion determinant as given in eq.(105). Again, starting from the general form of G and after quite lengthy calculations that we shall not reproduce here, one can find a unique solution for V_μ and $H_{\nu\alpha}$ leading to a gauge invariant action,

$$V^{(1)}_\mu[A] = \pm\frac{2i}{3m}{}^*f_\mu[A] \tag{114}$$

$$G^{(1)}[A,b,h,\bar{c}] = G^{(0)}[A,b,h,\bar{c}] \mp \frac{1}{96\pi m}tr\int d^3x\,\bar{c}_\mu\varepsilon_{\mu\nu\alpha}\varepsilon_{\nu\rho\sigma}$$
$$\Big(\frac{1}{2}\,[\,f_{\rho\sigma}[A-h] + 3f_{\rho\sigma}[b-h] - 2D_\rho[A-h](A_\sigma - b_\sigma)\,,\,(A_\alpha - b_\alpha)\,]$$
$$+\,4\,[\,f_{\rho\sigma}[A-h]\,,\,h_\alpha\,]\Big) \tag{115}$$

The corresponding change of variables (97) takes now the form

$$b_\mu = 2b'_\mu - A_\mu \pm \frac{2i}{3m}{}^*f_\mu[A] \tag{116}$$

and the decoupled effective action reads

$$S^{(1)}_{eff}[b,A,\bar{c},c] = S^{(0)}_{eff}[b,A,\bar{c},c] + tr\int d^3x\,\left(\frac{1}{6\pi m}f^2_{\mu\nu}[b'] + \frac{1}{24\pi m}f^2_{\mu\nu}[A]\right) \tag{117}$$

so that one can again integrate out the completely decoupled ghosts and b' fields ending with the bosonic action

$$S_{bos}^{(1)}[A] = \pm \frac{i}{16\pi} S_{CS}[A] + \frac{1}{24\pi m} tr \int d^3x\, f_{\mu\nu}^2[A] \tag{118}$$

This result extends to order $1/m$ the bosonization recipe presented in refs. [10,14].

In this way, from the knowledge of the $1/m$ expansion of the fermion determinant one can systematically find order by order the decoupling change of variables and construct the corresponding action for the bosonic field A_μ. One finds for the change of variables

$$b_\mu = 2b'_\mu - A_\mu \pm \frac{2i}{3m} {}^*f_\mu[A] + \frac{1}{m^2} C^{(2)} D_\rho[A] f_{\mu\rho}[A] + \ldots \tag{119}$$

Here $C^{(2)}$ is a (dimensionless) constant to be determined from the $1/m^2$ term in the fermion determinant expansion, which should be proportional to ${}^*f_\mu D_\rho f_{\rho\mu}$. Evidently, finding the BRST exact form becomes more and more involved and so is the form of the bosonic action which however, can be compactly written as

$$S_{bos}[A] = tr \int d^3x \, \Big(L[-A + V[A]]$$

$$\mp \frac{i}{16\pi} (2A_\mu - V_\mu[A]) ({}^*f_\mu[-A + V[A]] + {}^*H_\mu[-A + V[A], A, 0]) \Big) \tag{120}$$

Let us end this section by writing the bosonization recipe for the fermion current accompanying this result for the bosonic action. From eq.(26) we have, in $d = 3$

$$\bar\psi^i \gamma_\mu t^a_{ij} \psi^j \to \pm \frac{i}{8\pi} \varepsilon_{\mu\nu\alpha} \partial_\nu A^a_\alpha \tag{121}$$

7 $2+1$ Fermions at Finite Temperature

The results above correspond to Quantum Field Theory at zero temperature. What about $T \neq 0$? Concerning the $2 + 1$ fermion determinant this question was first addressed in [43] where it was argued that the coefficient of the induced CS term remains unchanged at finite temperature. Contrasting with this analysis, perturbative calculations yielded effective actions with CS coefficients which are smooth functions of the temperature [44-52]. It is important to notice that these computations dealt with the fermion mass dependent parity breaking and ignored the parity anomaly related to gauge invariant regularizations.

The issue of renormalization of the CS coefficient induced by fermions at $T \neq 0$ was reanalysed in refs.[53-54] where it was concluded that, in perturbation theory and on gauge invariance grounds, the effective action for the gauge field cannot contain the smoothly renormalized CS coefficient which was the answer of perturbative calculations. More recently, the exact result for the effective action of a $0+1$ massive fermion system [55] as well as zeta function calculations of the effective action in the $2+1$ Abelian case [56] and its explicit temperature dependent parity breaking part in a particular background [57-58] have explicitly shown that although the perturbative expansion leads to a non-quantized T-dependent CS coefficient, the complete effective action can be seen to be gauge invariant under both small and large gauge transformations, the temperature depending shift in the CS coefficient being just a byproduct of considering just the first term in the expansion of the effective action.

There is by now an extended literature on the issue of finite temperature and CS terms, mainly discussing abelian examples. To our knowledge, only in [58] a non-Abelian case is discussed and we shall then concentrate in this last calculation referring to [55-66] for details on the abelian case.

We are concerned with

$$\Gamma_{odd}(A,M) = \frac{1}{2}(\Gamma(A,M) - \Gamma(A,-M)) \tag{122}$$

where

$$\exp\left(-\Gamma(A,M)\right) = \int \mathcal{D}\psi\,\mathcal{D}\bar{\psi}\ \exp\left[-S_F(A,M)\right], \tag{123}$$

and $S_F(A,M)$ is the action for massive fermions (with mass M) in a gauge background A_μ. As mentioned above (see [56] for a discussion), the mass dependent parity violating term is not the only one arising in $\Gamma(A,M)$; there is also a local parity anomaly contribution in the form of half a CS term arising in any gauge invariant regularization. This term, first noticed at $T=0$ in [8] for massless fermions and in [31] for massive fermions, is mass and temperature independent and can be removed by a local counterterm at the price of breaking large gauge invariance. It is in fact not taken into account in most of the literature analysing $2+1$ dimensional massive fermion models. To understand the interplay between the two contributions, one can regard the mass dependent parity violating term as naturally arising due to the fact that the Lagrangian already contains at the classical level a parity violating mass term. Concerning the mass-independent contribution which comes from the parity anomaly, it can be seen as a necessary consequence of any gauge invariant regularization of the path integral fermionic measure. After these remarks, it is clear that our definition of Γ_{odd} excludes this last anomalous contribution but since it is temperature independent, it does not affect our analysis.

The calculation of (122) for the general case, namely, for *any* gauge field configuration is not something we can do exactly. Instead of making a perturbative calculation dealing with a small but otherwise arbitrary gauge field configuration, we shall consider a restricted set of gauge field configurations which can however be treated exactly.

In order to get an exact result we choose a particular gauge field background which corresponds to a vanishing color electric field and a time-independent color magnetic field,

$$A_3 = A_3(\tau), \tag{124}$$

$$A_j = A_j(x) \;\; (j = 1, 2) \tag{125}$$

or any equivalent configuration by gauge transformations. In the non-Abelian case we further restrict A_3 to point in a fixed direction in the internal space,

$$A_3 = |A_3|\check{n}, \tag{126}$$

and A_j to commute with A_3,

$$[A_j, A_3] = 0 \;\; (j = 1, 2)\,. \tag{127}$$

Although for $SU(2)$ this implies that all of the components of A_μ commute, and can be thus seen as an "Abelian-like" configuration, for $SU(N)$ with $N > 2$ one can see that genuine non-Abelian effects are incorporated. The configurations under consideration are reminiscent of the ones treated in [67] for massless fermions at $T = 0$; in that case Lorentz covariance of the local result allowed straightforward generalization to arbitrary backgrounds. Unfortunately, this will not be the case here.

Our main result can be presented through the formula we obtain for Γ_{odd},

$$\Gamma_{odd} \;=\; \frac{ig}{4\pi} tr \left(\arctan[\tanh(\frac{\beta M}{2}) \tan(\frac{g}{2} \int_0^\beta A_3 d\tau)] \int d^2x \, \epsilon_{ij} F_{ij} \right) \tag{128}$$

where g is the coupling constant, $\beta = 1/T$ and tr is an adequate trace in $SU(N)$ (matrix functions are defined as usual as power series).

The Euclidean fermionic action which describes the system is written as

$$S_F(A, M) \;=\; \int_0^\beta d\tau \int d^2x \, \bar{\psi}(\not{D} + M)\psi \,, \tag{129}$$

where the covariant derivative acting on the fermions in the fundamental representation of $SU(N)$ is defined as

$$D_\mu = \partial_\mu + ig\, A_\mu \,, \tag{130}$$

and the gauge connection A_μ is written as

$$A_\mu \;=\; A_\mu^a \, \tau_a \tag{131}$$

with τ_a denoting hermitian generators of the Lie algebra ($a = 1, \ldots, N^2 - 1$), verifying the relations

$$\tau_a^\dagger \;=\; \tau_a, \quad [\tau_a, \tau_b] \;=\; i f_{abc} \tau_c, \quad \mathrm{tr}(\tau_a \tau_b) = \frac{1}{2} \delta_{ab} \,, \tag{132}$$

with f_{abc} the totally antisymmetric structure constants. For the particular case of $SU(2)$, which we shall consider in more detail, we have $f_{abc} = \epsilon_{abc}$ since the generators will be taken to be the usual Pauli matrices.

We are concerned with the parity-odd piece of the efective action defined in (122). Fermionic (bosonic) fields satisfy antiperiodic (periodic) boundary conditions in the timelike direction.

We shall in this case restrict the set of configurations for the gauge fields given by (124)-(127) in order to be able to calculate Γ_{odd} exactly. Before doing so, let us clarify a point about the nature of the gauge group boundary conditions in imaginary time.

Non-Abelian gauge transformations are defined by their action on the fermionic and gauge fields

$$\begin{aligned}
\psi(\tau, x) \;\to\; \psi^U(\tau, x) \;&=\; U(\tau, x)\, \psi(\tau, x), \\
\bar{\psi}(\tau, x) \;\to\; \bar{\psi}^U(\tau, x) \;&=\; \psi(\tau, x)\, U^\dagger(\tau, x)
\end{aligned}$$

$$A_\mu(\tau, x) \;\to\; A_\mu^U(\tau, x) \;=\; U(\tau, x) A_\mu(\tau, x) U^\dagger(\tau, x) \;-\; \frac{i}{g} U(\tau, x) \partial_\mu U^\dagger(\tau, x) \,. \tag{133}$$

In order to decide the boundary conditions the gauge group element should satisfy in the timelike direction, one requires that the periodicity of the gauge field and the antiperiodicity of the fermions is unaltered under a gauge transformation. Concerning the gauge field, this only imposes on U the condition

$$U(\beta, x) \;=\; h\, U(0, x) \tag{134}$$

where h is an element of Z_N, the center of $SU(N)$. Now, concerning fermions, the condition on U depends on whether they are in the fundamental or adjoint representation. In the fundamental one, it is easily seen that

$$U(\beta, x) \;=\; U(0, x) \tag{135}$$

while in the adjoint representation, condition (134) follows instead. As we assume fermions are in the fundamental representation, the group elements $U(\tau, x)$ are taken to be strictly periodic. One can then prove [69] that for compact groups

$$w(U) = \frac{1}{12\pi^2 N} tr \int d^3x \epsilon_{\mu\nu\alpha} U^{-1}\partial_\mu U U^{-1}\partial_\nu U U^{-1}\partial_\alpha U \qquad (136)$$

is an integer number that labels homotopically equivalent gauge transformations. Thus the disctintion between large and small gauge transformations has a different origin here than in the Abelian case.

We thus consider a class of configurations equivalent by gauge transformations to

$$A_3 = |A_3|(\tau)\check{n}, \qquad (137)$$

$$A_j = A_j(x)\,,\ [A_j, \check{n}] = 0\ \ (j = 1,2)\,. \qquad (138)$$

where $\check{n}$ is a fixed direction in the Lie algebra ($\check{n} = n^a\tau_a$, $n^a n^a = 1$).

We note that conditions (137)-(138) assure the vanishing of the colour electric fields, as well as the time independence of the colour magnetic fields. Regarding the condition (138), which requires the spatial gauge field components to commute with A_3, it is worth remarking that its consequences depend strongly on whether the group considered is $SU(2)$ or $SU(N)$ with $N > 2$. In the former case, the only solution to (138) corresponds to a configuration with all the gauge field components pointing in the same direction $\check{n}$ in internal space, i.e. an 'Abelian like' configuration. In contrast, for $N > 2$, configurations with $[A_1, A_2] \neq 0$ are indeed possible.

To make the point above more explicit let us analyse the simple specific example of $SU(3)$ with the generators given by the standard Gell-Mann matrices; one can then take A_1 and A_2 as linear combinations of τ_1, τ_2 and τ_3 (generators of a $SU(2)$ subgroup) and A_3 pointing in the direction of τ_8. This situation easily generalizes to $N > 3$ since one can construct the set of generators for a higher N in such a way that it contains the generators corresponding to $SU(N-1)$ as a subset of block-diagonal matrices, and one of the extra generators can be always defined as to commute with them. Thus it is possible to take A_1 and A_2 as non commuting vectors in the subalgebra corresponding to $SU(N-1)$ and A_3 commuting with them.

Coming back to the general case, let us point that, as in the Abelian case, one can erase the τ dependence of A_3 component by considering a change of variables for the fermionic fields corresponding to a gauge transformation of the form

$$U(t) = e^{ig\Omega(\tau)\check{n}} \qquad (139)$$

and

$$\Omega(\tau) = -\int_0^\tau d\tau' A_3^{(\check{n})}(\tau') + \left(\frac{1}{\beta}\int_0^\beta d\tau' A_3^{(\check{n})}(\tau')\right)\tau. \tag{140}$$

Now, because of condition (138) the space components of the gauge field remain unchanged under this transformation, while A_3 takes the constant value $\tilde{A}_3 = \frac{1}{\beta}\int_0^\beta d\tau A_3(\tau) = |\tilde{A}_3|\check{n}$. After these remarks, we assume a gauge transformation has been made on the fermions in order to reach a constant $\tilde{A}_3$ and the rest of conditions (137)-(138) for the gauge field.

After a Fourier transformation on the time variable for ψ and $\bar{\psi}$ of the form

$$\psi(\tau, x) = \frac{1}{\beta}\sum_{n=-\infty}^{+\infty} e^{i\omega_n\tau}\psi_n(x)$$

$$\bar{\psi}(\tau, x) = \frac{1}{\beta}\sum_{n=-\infty}^{+\infty} e^{-i\omega_n\tau}\bar{\psi}_n(x)\,, \tag{141}$$

the Euclidean action can be written as an infinite series of decoupled actions,

$$S_F = \frac{1}{\beta}\sum_{n=-\infty}^{+\infty}\int d^2x\bar{\psi}_n(x)\left[\not{d} + M + i\gamma_3(\omega_n + g\tilde{A}_3^a\tau_a)\right]\psi_n(x) \tag{142}$$

where $\not{d} = \gamma_j(\partial_j + igA_j)$ is the non-Abelian Dirac operator corresponding to the spatial coordinates and the spatial components of the gauge field. Concerning the fermionic measure, we write it in the form

$$\mathcal{D}\psi(\tau, x)\,\mathcal{D}\bar{\psi}(\tau, x) = \prod_{n=-\infty}^{n=+\infty}\mathcal{D}\psi_n(x)\,\mathcal{D}\bar{\psi}_n(x)\,, \tag{143}$$

so that again the $2+1$ determinant becomes an infinite product of the corresponding $1+1$ Euclidean Dirac operators

$$\det(\not{\partial} + ig\not{A} + M) = \prod_{n=-\infty}^{n=+\infty}\det[\not{d} + M + i\gamma_3(\omega_n + g\tilde{A}_3^a\tau_a)]\,. \tag{144}$$

We now show that one can decouple of parity breaking and parity conserving parts of the determinant First, we use the property

$$M + i\gamma_3(\omega_n + g\tilde{A}_3^a\tau_a) = \rho_n\, e^{i\phi_n} \tag{145}$$

where

$$\rho_n = \sqrt{M^2 + (\omega_n + g\tilde{A}_3^a\tau_a)^2}\,;\, \phi_n = \arctan(\frac{\omega_n + g\tilde{A}_3^a\tau_a}{M})\,. \tag{146}$$

The usual definition of functions of matrices in terms of power series has been used above. It is important to realize that, being ϕ_n a non-trivial Hermitean function of a matrix in the Lie algebra, it will in general have components along the generators τ_a and also along the identity matrix, namely,

$$\phi_n = \phi_n^0 1 + \phi_n^a \tau_a . \tag{147}$$

As an illustration, we consider the $SU(2)$ case. A somewhat lenghty but otherwise straightforward calculation yields explicit expressions for these components of ϕ_n:

$$\begin{aligned} \phi_n^0 &= \frac{1}{2} \arctan\left(\frac{2M\omega_n}{M^2 + \frac{g^2}{4}|\tilde{A}_3|^2 - \omega_n^2} \right) \\ \phi_n^a &= \arctan\left(\frac{gM|\tilde{A}_3|}{M^2 - \frac{g^2}{4}|\tilde{A}_3|^2 + \omega_n^2} \right) n^a . \end{aligned} \tag{148}$$

The $1+1$ determinant for a given mode is a functional integral over $1+1$ fermions that using (145) can be written as

$$\det[\not{d} + M + i\gamma_3(\omega_n + g\tilde{A}_3^a\tau_a)] =$$
$$\int \mathcal{D}\chi_n \mathcal{D}\bar{\chi}_n \exp\left\{ - \int d^2x \bar{\chi}_n(x)(\not{d} + \rho_n e^{i\gamma_3\phi_n})\chi_n(x) \right\} .$$

We now perform the change of fermionic variables

$$\chi_n(x) = e^{-i\frac{\phi_n}{2}\gamma_3}\chi'_n(x) \; , \; \bar{\chi}_n(x) = \bar{\chi}'_n(x)e^{-i\frac{\phi_n}{2}\gamma_3} , \tag{149}$$

and verify that due to the last condition in (138) it indeed decouples the parity violating part of the effective action. We find, including the anomalous Fujikawa Jacobian

$$\det[\not{d} + M + ig\gamma_3(\omega_n + \tilde{A}_3^a\tau_a)] = J_n \det[\not{d} + \rho_n]. \tag{150}$$

The Jacobian in (150) reads [68]

$$J_n[A, M] = \exp\left[-itr\frac{\phi_n}{2} \int d^2x \mathcal{A} \right] , \tag{151}$$

with $\mathcal{A} = \mathcal{A}^a\tau^a$ denoting the $1+1$ Euclidean anomaly under an infinitesimal non-Abelian axial transformation. As this transformation is x-independent, there is no difference between finite and infinitesimal transformations and one can just simply iterate the infinitesimal Fujikawa Jacobian [68] in order to get the finite answer (151). Also note that ϕ_n^0 (the component along the identity) does not contribute to the jacobian since $tr(\phi_n^0\mathcal{A}) = 0$. A standard calculation leads for the two-dimensional non-abelian anomaly the answer (see for example [70])

$$\mathcal{A} = \frac{g}{2\pi}\epsilon_{ij}F_{ij} \tag{152}$$

so that the Jacobian finally takes the form

$$J_n[A, M] = \exp\left[-\frac{ig}{4\pi}tr\left(\phi_n \int d^2x\, \epsilon_{ij}F_{ij}\right)\right] . \tag{153}$$

We see from eqs.(144) and (150) that the parity odd piece of the effective action is again given in terms of the infinite set of n-dependent Jacobians,

$$\Gamma_{odd}[A, M] = -\sum_{n=-\infty}^{n=+\infty} \log J_n[A, M] = \frac{ig}{4\pi}tr\left(\left(\sum_{n=-\infty}^{+\infty}\phi_n\right)\int d^2x\, \epsilon_{ij}F_{ij}\right) . \tag{154}$$

Now we have to perform the summation over the ϕ_n's. As explicitly shown in [58], the answer is

$$\Gamma_{odd} = \frac{ig}{4\pi}tr\left(\arctan[\tanh(\frac{\beta M}{2})\tan(\frac{g}{2}\beta\tilde{A}_3)]\int d^2x\, \epsilon_{ij}F_{ij}\right) . \tag{155}$$

We can check this result by doing explicit computations with the components ϕ_n^a given in eq.(148) for the $SU(2)$ case. From eq.(154),

$$\Gamma_{odd}[A, M] = \frac{ig}{8\pi}\sum_{n=-\infty}^{+\infty}\phi_n^a \int d^2x\, \epsilon_{ij}F_{ij}^a. \tag{156}$$

Using eq. (148) we have to compute

$$\Sigma = \sum_{n=-\infty}^{\infty} \arctan\left(\frac{gM|\tilde{A}_3|}{M^2 - \frac{g^2}{4}|\tilde{A}_3|^2 + \omega_n^2}\right) , \tag{157}$$

or, in terms of dimensionless variables

$$m = \beta M \qquad x = \frac{g}{2}\beta|\tilde{A}_3| \tag{158}$$

$$\Sigma(x, m) = \sum_{n=-\infty}^{\infty} \arctan\left(\frac{2mx}{m^2 - x^2 + (2n+1)^2\pi^2}\right) . \tag{159}$$

The sum is convergent, but in order to calculate Σ it will be convenient to write

$$\Sigma(x, m) = \int_0^x du \frac{\partial\Sigma}{\partial u}(u, m). \tag{160}$$

The implicit subtraction of a zero-field contribution vanishes term by term in this case (in contrast with what happens in the abelian case [57]).

After some calculations, one has

$$\frac{\partial \Sigma}{\partial x}(x,m) = 2m \sum_{n=-\infty}^{\infty} \frac{m^2 + (2n+1)^2\pi^2 + x^2}{[m^2 + (2n+1)^2\pi^2 - x^2]^2 + 4m^2x^2} \tag{161}$$

One could now arrange this expression using a standard Regge-type trick to rewrite (161) as a contour integral of the form

$$\frac{\partial \Sigma}{\partial x}(x,m) = -\frac{m}{2\pi i} \oint_C dz \tanh(z/2) \frac{m^2 - z^2 + x^2}{[m^2 - z^2 - x^2]^2 + 4m^2x^2} \tag{162}$$

where C is a contour including all the poles of $\tanh(z/2)$. After continuing C into the upper and lower half-planes to pick up the 4 poles of the fraction only, we end with

$$\frac{\partial \Sigma}{\partial x}(x,m) = \frac{i}{2}[\tanh(\frac{x - im}{2}) - \tanh(\frac{x + im}{2})] \tag{163}$$

Using this expression in (160) we finally get

$$\Sigma(x,m) = 2\arctan[\tanh(m/2)\tan(x/2)] \tag{164}$$

so that Γ_{odd} can be written as

$$\Gamma_{odd} = \frac{ig}{4\pi} \arctan[\tanh(\frac{\beta M}{2})\tan(\frac{g}{4}\beta|\tilde{A}_3|)]n^a \int d^2x\, \epsilon_{ij} F^a_{ij}. \tag{165}$$

Finally, observing that $(n^a\tau_a)^{(2k+1)} = \frac{1}{2^{2k}} n^a\tau_a$ and only odd powers enter the expansions of the functions involved, we see that the result is identical to eq.(155).

In order to analyze the result (155) let us write it in the most general form

$$\Gamma_{odd} = \frac{ig}{4\pi} tr \left(\arctan[\tanh(\frac{\beta M}{2})\tan(\frac{g}{2}\int_0^\beta d\tau A_3(\tau))] \int d^2x \epsilon_{ij} F_{ij} \right) \tag{166}$$

Then we note that in the zero-temperature limit one has

$$\lim_{T\to 0} \Gamma_{odd} = \frac{ig^2}{8\pi}\frac{M}{|M|} tr \left(\int_0^\beta d\tau A_3(\tau) \int d^2x\, \epsilon_{ij} F_{ij} \right). \tag{167}$$

This result is the usual one, namely

$$\lim_{T\to 0} \Gamma_{odd} = \frac{1}{2}\frac{M}{|M|} S_{CS}, \tag{168}$$

restricted to the particular background we have considered. Here S_{CS} is the non-Abelian CS action

$$S_{CS} = \frac{ig^2}{8\pi} \int d^3x \epsilon_{\mu\nu\alpha} tr(F_{\mu\nu} A_\alpha - \frac{2}{3} A_\mu A_\nu A_\alpha) \tag{169}$$

which for a gauge field satisfying the restrictions (138) reads

$$S_{CS} = \frac{ig^2}{4\pi} tr \int d^3x \, A_3 \epsilon_{ij} F_{ij}. \tag{170}$$

We thus recover the zero-temperature result first obtained in [8] by calculating the v.e.v. of the fermion current in a constant non-Abelian field strength tensor background or in [67] in a static non-Abelian magnetic background like ours. We recall, however, that gauge invariance under large gauge transformations is obtained only when the parity anomaly $\pm\frac{1}{2}S_{CS}$ is added to the mass- and temperature-dependent expression for Γ_{odd}.

We finally note that a perturbative expansion in powers of the coupling constant g shows a smooth temperature dependence of the CS coefficient,

$$\Gamma_{odd} = \frac{1}{2} \tanh(\frac{M\beta}{2}) S_{CS} + O(e^4). \tag{171}$$

Concerning gauge invariance of the finite temperature result we note that, in contrast to the Abelian case, there is no room for large gauge transformations preserving the conditions (137) and (138) under wich our result (166) was obtained. We can only quote gauge invariance under small gauge transformations that do not mix spatial and time components. However, we expect that the large gauge invariance apparently broken by the perturbative expansion (171) should be recovered by the full result.

Acknowledgements

I would like to thank Claudio Teitelboim and Jorge Zanelli for the organization of the Meeting and all the people from Base Aérea Antártica Pdte. E. Frei M. for their splendid reception in Antartida. F.A.S. is partially supported by CONICET and CICBA, Argentina and a Commission of the European Communities contract No: C11*-CT93-0315.

References

1. P.B. Wiegmann, Phys. Rev. Lett. **60** (1987) 821; I. Dzyaloshinski, A.M. Polyakov and P.B. Wiegmann, Phys. Lett. **127A** (1988) 112.
2. A.M. Polyakov, Mod. Phys. Lett. **A3**, 325 (1988).

3. See for example E. Fradkin, *Field theories of Condensed Matter Physics*, Frontiers in Physics, New York, 1991 and references therein.
4. J.F. Schonfeld, Nucl. Phys. **B185**, 157 (1981).
5. S. Deser, R. Jackiw and S. Templeton, Phys. Rev. Lett. **48**, 975 (1982); Ann. of Physics (N.Y) **140**, 372 (1982).
6. A.S. Schwartz, Mod. Phys. Lett. **2**, 247 (1978).
7. E. Witten, Comm. Math. Phys. **121**, 351 (1989).
8. A.N. Redlich, Phys. Rev. Lett. **52**, 18 (1984); Phys. Rev. **D29**, 236 (1984).
9. E. Fradkin and F.A. Schaposnik, Phys. Lett. **B338**, 253 (1994).
10. N. Bralić, E. Fradkin, M.V. Manías and F.A. Schaposnik, Nucl. Phys. **B446**, 144 (1995).
11. F.A. Schaposnik, Phys. Lett. **B356**, 39 (1995).
12. J.C. Le Guillou, C. Núñez, and F.A. Schaposnik, Ann. of Phys. (N.Y.) **251**, 426 (1996).
13. C.D. Fosco and F.A. Schaposnik, Phys. Lett. **B391**, 136 (1997).
14. J.C. Le Guillou, E. Moreno, C. Núñez and F.A. Schaposnik,Nucl. Phys. **B484**, 682 (1997).
15. J.C. Le Guillou, E. Moreno, C. Núñez and F.A. Schaposnik, Phys. Lett. **B409** (1997) 257
16. C.P. Burgess, C.A. Lütken and F. Quevedo, Phys. Lett. **B336**, 18 (1994).
17. P.A. Marchetti in *Common Trends in Condensed Matter and High Enery Physics, Chia Laguna, Italy,* Sep. 1995. hep-th/9511100 report.
18. C.P. Burgess and F. Quevedo, Phys. Lett. **B329**, 457 (1994).
19. J.L. Cortés, E. Rivas, and L. Velázquez, Phys. Rev. **D53**, 5952 (1996).
20. F.A. Schaposnik, *Bosonization in $d > 3$ dimensions*, Proceedings of the CERN-Santiago de Compostela-La Plata Meeting, eds. H. Falomir et al, AIP Press 1997.
21. A. Luther, Phys. Rev. **D19**, 320 (1979).
22. A.M. Polyakov, Mod. Phys. Lett. **A3**, 325 (1988).
23. M. Luscher, Nucl. Phys. **B326**, 557 (1989).
24. F.D.M. Haldane, Helv. Phys. Acta **65**, 52 (1992).
25. A. Kovner and P.S. Kurzepa, Phys. Lett. **B321**, 129 (1994).
26. E.C. Marino, Phys. Lett. **B263**, 63 (1991).
27. J. Frohlich, R. Götschmann and P.A. Marchetti, J. Phys. **A28**, 1169 (1995) .
28. D.G. Barci, C.D. Fosco and L.E. Oxman, Phys. Lett. **B375**, 267 (1996).
29. P.H. Damgaard, H.B. Nielsen and R. Sollacher, Nucl. Phys. **B385**, 227 (1992); Phys. Lett. **B296**, 132 (1992).
30. P.H. Damgaard, F. De Jonghe and R. Sollacher, Nucl. Phys. **B454**, 701

(1995).
31. R.E. Gamboa Saraví, G.L. Rossini and F.A. Schaposnik, Int. J. Mod. Phys. **A11** 2643 (1996).
32. P.K. Townsend, K. Pilch and P. van Nieuwenhuizen, Phys. Lett. **B136**, 38 (1984); **B137**, 443 (1984).
33. S. Deser S. and R. Jackiw, Phys. Lett. **B139**, 371 (1984).
34. I.J.R. Aitchison, C.D. Fosco, and J. Zuk, Phys. Rev. **D48**, 5895 (1993).
35. D. Gross in *Methods in Field Theory*, Eds. R. Balian and J. Zinn-Justinn, North-Holland, 1976.
36. See for example S. Coleman, *Aspects of Symmetry*, Cambridge University Press, Cambridge, 1985.
37. J.D. Bjorken, Phys. Rev. **148**, 1467 (1966).
38. K. Johnson and F.E. Low, Prog. Theoret. Phys. (Kyoto), Suppl. **37-38**, 74 (1966).
39. For a review see. e.g. R. Jackiw, R. *Lectures on Current Alebra and its Applications*, (eds. Treiman, S.B., Jackiw, R. and Gross, D.J.), Princeton University Press, 1972.
40. D.G. Boulware and R. Jackiw, Phys. Rev. **186**, 1442 (1969).
41. H.J. de Vega and H.O. Girotti, Nucl. Phys. **B 79**, 77 (1974).
42. M.S. Chanowitz, Phys. Rev. **D 2**, 3016 (1970); **D 4**, 1717 (1971).
43. R. Pisarski, Phys. Rev. **D35**, 664 (1987).
44. A.J. Niemi and G.W. Semenoff, Phys. Rev. Lett. **51**, 2077 (1983) .
45. A.J. Niemi, Nucl. Phys. **B251**, 55 (1985).
46. A.J. Niemi and G.W. Semenoff, Phys. Rep. **135**, 99 (1986).
47. K. Babu, A. Das and P. Panigrahi, Phys. Rev. **D36**, 3725 (1987).
48. A. Das and S. Panda, J. Phys. A: Math. Gen. **25**,L245 (1992) .
49. E.R. Poppitz, Phys. Lett. **B252**, 417 (1990).
50. M. Burgess, Phys. Rev. **D44**, 2552 (1991) .
51. W.T. Kim, Y.J. Park, K.Y. Kim and Y. Kim, Phys. Rev. **D 46**, 3674 (1993).
52. K. Ishikawa and T. Matsuyama, Nucl. Phys. **B 280** [F518], 523 (1987).
53. N. Bralić, C.D. Fosco and F.A. Schaposnik, Phys. Lett. **B 383** 199 (1996).
54. D. Cabra, E. Fradkin, G. Rossini and F.A. Schaposnik, Phys. Lett. B 383, 434 (1996).
55. G. Dunne, K. Lee, and Ch. Lu, Phys. Rev. Lett. **78**, 3434 (1997).
56. S. Deser, L. Griguolo and D. Seminara, Phys. Rev. Lett. **79** (1997) 1976.
57. C.D. Fosco, G. Rossini and F.A. Schaposnik, Phys. Rev. Lett. **79** (1997) 1980; *Erratum* ibid **79** (1997) 4296.
58. C.D. Fosco, G. Rossini and F.A. Schaposnik, Phys. Rev. **D56** (1997)

6547.
59. I. Aitchinson and C. Fosco, Phys. Rev. **D57** (1998) 1171.
60. S. Deser, L. Griguolo, and D. Seminara, Phys. Rev. **D57**, 7444 (1998).
61. A. Das and G. Dunne, Phys. Rev. **D57**, 5023 (1998).
62. S. Deser, L. Griguolo and D. Seminara, Commun. Math. Phys. **197** 443 (1998).
63. R. González Felipe, Phys. Lett. **B417** (1998) 114.
64. L.L. Salcedo, hep-th/9802071.
65. R. Jackiw and S.Y. Pi, Phys. Lett. **B423** 364 (1998).
66. S. Deser, G. Dunne, L. Griguolo, C. Lu, K. Lee and D. Seminara, hep-th/9802075.
67. A.J. Niemi and G.W. Semenoff, Phys. Rev. Lett. **51**, 2077 (1983) .
68. K. Fujikawa, Phys. Rev. Lett. **42** 1195 (1979); Phys. Rev. **D21**, 2848 (1980).
69. O.Alvarez, Commun. Math. Phys. **100** (1985) 279.
70. R.E. Gamboa Saraví, F.A. Schaposnik and J.E. Solomin, Nucl.Phys. **B185** (1981) 239.

UNIQUENESS OF D=11 SUPERGRAVITY

S. DESER
Department of Physics
Brandeis University
Waltham, MA 02254, USA

We study the extent to which D=11 supergravity can be deformed and show in two very different ways that, unlike lower D versions, it forbids an extension with cosmological constant. Some speculations about other invariants are made, in connection with the possible counterterms of the theory.

It is a pleasure to report here on work done jointly with K. Bautier, M. Henneaux, and D. Seminara, one of whom is closely connected with this Center and all of whom I thank for their contribution to this paper. The completed aspects of our research have also just appeared in print [1].

Anyone studying supergravities cannot fail to marvel at how the interplay of Lorentz invariance, Clifford algebra and gauge field properties conspire to limit their dimensionality. In particular not only is D=11 maximal if gravity is to remain the highest spin of the supermultiplet, but only one configuration of fields, the N=1 graviton plus vector-spinor and 3-form potential combination is permitted[a][2]. There is some "fine print" involved as well. For example, I will be talking solely about actions whose gravitational component is Einstein, rather than say Chern–Simons – where very recent work at this Center [4] has revealed other supersymmetric D=11 possibilities. Beyond D=11, one would require the appearance of spin >2 fields and/or more than one graviton, both of which are known [5,6] to lead to inconsistencies.

Despite the magic of D=11, the theory was for many years neglected (if never forgotten – as an inspiration for KK descents to lower dimensions, if nothing else), because superstrings had their own magic number, D=10. Faith in the importance of D=11 was revived when it was seen to be the low energy sector of M-theory unification, and that "dimensional enhancement" was as interesting as dimensional reduction. So this is a good time to understand more deeply just how unique a theory it really is – within the framework I have indicated, requiring that it have an Einstein term to describe the graviton.

[a] For a quick counting, recall that graviton excitations are described by transverse traceless spatial components g_{ij}^{TT}, hence there are (D-2)(D-1)/2 -1 = D(D-3)/2; the spatial three-form A_{ijk}^{T} are also transverse hence (D-2)(D-3)(D-4)/3!, while the vector-spinors have (D-3) $2^{[D/2]-1}$ excitations.

Now there is one other question in physics that is on an equally mysterious footing as what is so special about D=11, namely why is Λ=0 – the cosmological constant problem. Early hopes that supersymmetry would solve it in the matter sector, through vacuum energy cancellations, were made to some extent irrelevant by the perfect consistency of cosmological constant extensions of supergravity. A cosmological term (of anti deSitter type) $\sim -|\Lambda|\sqrt{-g}$ could be added to a masslike term for the fermion $\sim \sqrt{|\Lambda|}\ \bar{\psi}_\mu \Gamma^{\mu\nu}\,\psi_\nu$ in a supersymmetric way, as was first realized for D=4 [7,8] and then extended all the way to D=10 [9]. Things got rather elaborate on the way up, with scalars for example decorating the cosmological term, but it was there. At D=11, however, there seemed to be a snag; surprisingly, there were only a couple of papers directed at this question at the time. To our knowledge, there have been two previous approaches to this result. One [3] consists in a classification of all graded algebras and consideration of their highest spin representations. Although we have not found an explicit exclusion of the cosmological extension in this literature, it is undoubtly implied there under similar assumptions. The second [10] considers the properties of a putative "minimal" graded Anti de Sitter algebra and shows it to be inconsistent in its simplest form. While one may construct generalized algebras that still contract to super-Poincaré, these can also be shown to fail, using for example some results of [11]. In [10], a Noether procedure, starting from the full theory of [2], was also attempted; as we shall show below, there is an underlying cohomological basis for that failure. A careful reconsideration of the problem, resulting in a no-go theorem is the main result to be reported here. To be sure, this does not exorcise the cosmological constant problem: it can reappear under dimensional reduction (as in fact discovered again recently [12]) as well of course as through supersymmetry breaking. Still, it should be appreciated that there is now one model–and a very relevant one it is–of a QFT that includes gravity and really excludes a Λ term, through the magic of supersymmetry. Although we have no "deep" physical selection rule to account for this, we can point to the mysterious 3-form field as the immediate cause. We also mention that current investigations of lower dimensional (brane) models also have a stake in the outcome (see eg. [13].)

We will proceed from two complementary starting points. The first will be the Noether current approach, in which we attempt—and fail!—to find a linearized, "globally" supersymmetric model about an Anti de Sitter (AdS) background upon which to construct a full locally supersymmetric theory. Since a Noether procedure is indeed a standard way to obtain the full theory, in lower dimensions, the absence of a starting point for it effectively forbids the extension. In contrast, the second procedure will begin with the full (original) theory of [2] and attempt, using cohomology techniques, to construct—also unsuccessfully—a consistent deformation of the model and of its transformation rules that would include the desired fermion mass term plus cosmological term extensions. In both cases, the obstruction is due to the 4– (or 7–) form field necessary to balance degrees of freedom.

First, we recall some general features relevant to the linearized approach. It is well-known that Einstein theory with cosmological term linearized about a background solution of constant curvature retains its gauge invariance and degree of freedom count, with the necessary modification that the vielbein field's gauge transformation is the background covariant $\delta h^a_\mu = D_\mu \xi^a$. Similarly it is also known that the free spin 3/2 field's gauge invariance in this space is no longer $\delta\psi_\mu = \partial_\mu \alpha(x)$ or even $D_\mu \alpha(x)$, but rather the extended form [8]

$$\delta\psi_\mu = \mathcal{D}_\mu\, \alpha(x) \equiv (D_\mu + m\gamma_\mu)\alpha(x) \tag{1}$$

where $\mathcal{D}_\mu$ has the property that $[\mathcal{D}_\mu, \mathcal{D}_\nu] = 0$ when the mass m is "tuned" to an AdS cosmological constant: $2m = \sqrt{-\Lambda}$ (in $D = 11$). The modified transformation (1) then keeps the degree of freedom count for ψ_μ the same as in flat space, provided—as is needed for consistency—that the ψ's action and field equations also involve $\mathcal{D}_\mu$ rather than D_μ. [This is of course the reason for the "mass" term $m\bar{\psi}_\mu \Gamma^{\mu\nu}\psi_\nu$ acquired by the spinor field to accompany the cosmological one for gravity.] Given the above facts, the 3-form potential $A_{\mu\nu\rho}$ still balances fermi/bose degrees of freedom here. [For now, we keep the same field content as in the flat limit.] Unlike the other two fields, its action only involves curls and so it neither needs nor can accomodate any extra terms in the background to retain its gauge invariance and excitation count; indeed, the only possible quadratic addition would be a – true – mass term $\sim \Lambda A^2$ that would destroy both (there would be 120, instead of the 84 massless, excitations). One can therefore expect, with reason, that the problem will lie in the form (rather than gravity) sector's transformation rules. In the AdS background, the desired "globally" supersymmetric free field starting point involves the Killing spinor $\epsilon(x)$, $\mathcal{D}_\mu \epsilon(x) = 0$, which is unrelated to the general gravitino gauge spinor $\alpha(x)$ in (1). [Note that we can neither use $\partial_\mu \epsilon = 0$ because space is curved, nor $D_\mu \epsilon = 0$ because only $\mathcal{D}_\mu$'s commute.] The rules are essentially fixed from the known flat background ones (to which they must reduce for $\Lambda = 0$),

$$\begin{aligned}
\delta\,\psi_\mu &= \delta_h \psi_\mu + \delta_A \psi_\mu \\
&= \left(\frac{1}{4} X_{\mu ab}(h)\Gamma^{ab} - m\gamma^a h_{\mu a}\right)\epsilon + i/144\,(\Gamma^{\alpha\beta\gamma\delta}{}_{\mu} - 8\,\Gamma^{\beta\gamma\delta}\delta^\alpha_\mu)\epsilon\, F_{\alpha\beta\gamma\delta}
\end{aligned}$$

$$\delta\, h_{\mu a} = -i\, \bar{\epsilon}\Gamma_a \psi_\mu \qquad\qquad \delta\, A_{\mu\nu\rho} = 3/2\, \bar{\epsilon}\Gamma_{[\mu\nu}\psi_{\rho]}. \tag{2}$$

The linearized connection $X(h)$ is derived by a linearized "vanishing torsion" condition $D_\mu h_{\nu a} + X_{\mu ab} e^b_\nu - (\nu\mu) = 0$; throughout, the background vielbein is $e_{\mu a}$ and its connection is $\omega_{\mu ab}(e)$. Now vary the spinorial action $I[\psi] = -1/2 \int (dx) \psi_\mu \Gamma^{\mu\alpha\beta} \mathcal{D}_\alpha \psi_\beta$ (world Γ indices are totally antisymmetric and $\Gamma^\mu = e^\mu{}_a \gamma^a$ etc.). It is easily checked that although $[\Gamma, \mathcal{D}] \neq 0$, varying $\bar{\psi}$ and ψ does yield the same contribution, and using (2) we find

$$\delta I[\psi] = \delta_h I[\psi] + \delta_A I[\psi] = -i/8 \int (dx) E^{\mu b}(-i\kappa\bar{\epsilon}\Gamma_a \psi_\mu)$$

$$- i/8 \int (dx)[D_\alpha F^{\alpha\mu\rho\sigma}(\bar{\epsilon}\Gamma_{[\mu\nu}\psi_{\rho]}) + m\bar{\psi}_\mu(\Gamma^{\mu\alpha\beta\rho\sigma} F_{\alpha\beta\rho\sigma})\epsilon] \,. \tag{3}$$

Here $E^{\mu b}$ is the variation of the Einstein cosmological action linearized about AdS. The form-dependent piece of (3) has a first part that behaves similarly, namely it is proportional to the form field action's variation $D_\alpha F^{\alpha\mu\rho\sigma}$ (the Chern–Simons term, being cubic, is absent at this level). With the transformation choice (2), the variation of the Einstein plus form actions almost cancels (3). There remains $\bar{\psi} F \epsilon$, the $A-$variation of the gravitino mass term. What possible deformations of the transformation rules (2) and of the actions might cancel this unwanted term? The only dimensionally allowed change in (2) is a term $\bar{\delta}\psi_\mu \sim m \not{A}_\mu \epsilon$; however, it will give rise to unwanted gauge-variant contributions from the $m\bar{\psi}\Gamma\psi$ term $\sim m^2 \bar{\psi}\Gamma A \epsilon$, that would in turn require a true mass term $I_m[A] \sim m^2 \int (dx) A^2$ to cancel, thereby altering the degree of freedom count. Indeed these two deformations, $\bar{\delta}\psi_\mu$ and $I_m[A]$, are the only ones that have nonsingular $m \to 0$ limits. A detailed calculation reveals, however, that even with these added terms, the action's invariance cannot be preserved. In particular, there are already variations of the A^2 term that cannot be compensated. A completely parallel calculation starting with a dual, 7-form, model yields precisely the same obstruction[b]: defining the 4–form dual of the 7–form, we have the same structure as the 4–form case, up to normalizations, and face the same non-cancellation problem; also here a mass term is useless.

Our second approach analyses the extension problem in the light of the master equation and its consistent deformations [15,16,17]; see [18] for a review of the master equation formalism appropriate to the subsequent cohomological considerations. One starts with the solution of the master equation $(S, S) = 0$

[b]The 7– form variant was originally considered by [14], who argued that it was excluded in the non-cosmological case, but the possibility for a cosmological extension was not entirely removed; the latter was considered and rejected at the Noether level in [10].

[18,19] for the action of an undeformed theory (for us that of [2]). One then tries to perturb it, $S \to S' = S + g\Delta S^{(1)} + g^2\Delta S^{(2)} + ...$, where g is the deformation parameter, in such a way that the deformed S' still fulfills the master equation $(S', S') = 0$. As explained in [15] any deformation of the action of a gauge theory and of its gauge symmetries, consistent in the sense that the new gauge transformations are indeed gauge symmetries of the new action, leads to a deformed solution S' of the master equation. Conversely, any deformation S' of the original solution S of the master equation defines a consistent deformation of the original gauge invariant action and of its gauge symmetries. In particular, the antifield–independent term in S' is the new, gauge-invariant action; the terms linear in the antifields conjugate to the classical fields define the new gauge transformations [15,20] while the other terms in S' contain information about the deformation of the gauge algebra and of the higher-order structure functions. To first order in g, $(S', S') = 0$ implies $(S, \Delta S^{(1)}) = 0$, *i.e.*, that $\Delta S^{(1)}$ (which has ghost number zero) should be an observable of the undeformed theory or equivalently $\Delta S^{(1)}$ is "BRST-invariant" - recall that the solution S of the master equation generates the BRST transformation in the antibracket. To second order in g, then, we have $(\Delta S^{(1)}, \Delta S^{(1)}) + 2(S, \Delta S^{(2)}) = 0$, so the antibracket of $\Delta S^{(1)}$ with itself should be the BRST variation of some $\Delta S^{(2)}$.

Let us start with the full nonlinear 4-dimensional $N = 1$ case, where a cosmological term *can* be added, for contrast with $D = 11$. The action is [21]

$$I_4[e^a_\mu, \psi_\lambda] = -\frac{1}{2}\int (dx)\big(\frac{1}{2}ee^{a\mu}e^{b\nu}R_{\mu\nu ab} + \overline{\psi}_\mu\Gamma^{\mu\sigma\nu}D_\sigma\psi_\nu\big), \tag{4}$$

where $e \equiv \det(e_{a\mu})$ and D_μ here is of course with respect to the full vierbein; it is invariant under the local supersymmetry (as well as diffeomorphism and local Lorentz) transformations

$$\delta e^a_\mu = -i\bar{\epsilon}\Gamma^a\psi_\mu\,, \quad \delta\psi_\lambda = D_\lambda\epsilon(x)\,, \tag{5}$$

and under those of the spin connection ω^{ab}_μ. The solution of the master equation takes the standard form

$$S = I_4 + \int\int (dx)(dy)\varphi^*_i(x)R^i_A(x,y)C^A(y) + X, \tag{6}$$

where the φ^*_i stand for all the antifields of antighost number one conjugate to the original (antighost number zero) fields $e_{a\mu}$, ψ_λ, and where the C^A stand for all the ghosts. The $R^i_A(x,y)$ are the coefficients of all the gauge transformations leaving I_4 invariant. The terms denoted by X are at least of antighost number two, i.e. contain at least two antifields φ^*_i or one of the antifields C^*_α conjugate to the ghosts. The quadratic terms in φ^*_i are also quadratic in the ghosts and arise because the gauge transformations do not close off-shell [22]. We next recall some cohomological background [18] related to the general solution of the "cocycle" condition $(S, A) \equiv sA = 0$ for A with zero ghost number. If one expands A in antighost number $A = A_0 + \bar{A}$, where $\bar{A}$ denotes antifield-dependent terms, one finds that the antifield-independent term A_0 should be on-shell gauge-invariant. Conversely, given an on-shell invariant function(al) A_0 of the fields, there is a unique, up to irrelevant ambiguity, solution A (the "BRST invariant extension" of A_0) that starts with A_0. Below we shall obtain the required A_0. The relevant property that makes the introduction of a cosmological term possible in four dimensions is the fact that a gravitino mass term $m \int (dx) e \overline{\psi}_\lambda \Gamma^{\lambda\rho} \psi_\rho$ defines an observable; one easily verifies that it is on-shell gauge invariant under (5). Hence, one may complete it with antifield-dependent terms, to define the initial deformation $m\Delta S^{(1)}$ that satisfies $(\Delta S^{(1)}, S) = 0$. The antifield-dependent contributions are fixed by the coefficients of the field equations in the gauge variation of the mass term. Specifically, since one must use the *undeformed* equations for the gravitino and the spin connection in order to verify the invariance of the mass term under supersymmetry transformations, these contributions will be of the form $\psi^* C$ and $\omega^* C$, where C is the commuting supersymmetry ghost. They then lead to the known [7] modification of the supersymmetry transformation rules for the gravitino and the spin connection when the mass term is turned on[c]. Having obtained an acceptable first order deformation, $m\Delta S^{(1)}$, we must in principle proceed to verify that $(\Delta S^{(1)}, \Delta S^{(1)})$ is the BRST variation of some $\Delta S^{(2)}$; indeed it is , with $\Delta S^{(2)} = 3/2 \int (dx) e$, as expected. There are no higher order terms in the deformation parameter m because the antibracket of $\Delta S^{(1)}$ with $\Delta S^{(2)}$ vanishes ($\Delta S^{(1)}$ does not contain the antifields conjugate to the vierbeins), so the complete solution of the master equation with cosmological constant is $S + m\Delta S^{(1)} + m^2 \Delta S^{(2)}$, the action of [7,8]. [The possibility of introducing the gravitino mass term as an observable deformation hinged on the availability of a dynamical curved geometry in the sense that while $(S, \Delta S^{(1)}) = 0$ is always satisfied, only then is $(\Delta S^{(1)}, \Delta S^{(1)})$ BRST exact, i.e. is there a second order –gravitational– deformation.]

To summarize the analysis of the four-dimensional case, we stress that the cosmological term appears, in the formulation without auxiliary fields followed here, as the second order term of a consistent deformation of the ordinary supergravity action whose first order term is the gravitino mass term, with

[c] A complete investigation of the BRST cohomology of $N = 1$ supergravity has been recently carried out in [23].

the mass as deformation parameter; it is completely fixed by the requirement that the deformation preserve the master equation and hence gauge invariance. This means, in particular, that the cosmological constant itself must be fine-tuned to the value $-4m^2$, as explained in [8].[d]

Let us now turn to the action I_{CJS} of [2] in $D = 11$. The solution of the master equation again takes the standard form[e]

$$S = I_{CJS} + \int\int (dx)(dy)\varphi_i^*(x)R_A^i(x,y)C^A(y) + \int (dx)C^{*\mu\nu}\partial_\mu\eta_\nu + \int (dx)\eta^{*\mu}\partial_\mu\rho + Z, \tag{7}$$

where the η_ν and ρ are the ghosts of ghosts and ghost of ghost of ghost necessary to account for the gauge symmetries of the 3-form $A_{\lambda\mu\nu}$, and where Z (like X in (6)) is determined from the terms written by the $(S,S) = 0$ requirement. As in $D = 4$, we seek a first-order deformation analogous to

$$\Delta S^{(1)} = \frac{1}{2}m\int (dx)e\overline{\psi}_\lambda\Gamma^{\lambda\rho}\psi_\rho \; + \; \text{antifield-dep.} \tag{8}$$

However, contrary to what happened at $D = 4$, the mass term no longer defines an observable, as its variation under local supersymmetry transformations reads

$$\delta\big(e\overline{\psi}_\lambda\Gamma^{\lambda\rho}\psi_\rho\big) \approx -\frac{i}{18}\overline{\psi}_\mu\Gamma^{\mu\alpha\beta\gamma\delta}\epsilon F_{\alpha\beta\gamma\delta} + O(\psi^3) \tag{9}$$

where $\approx$ means equal on shell up to a divergence. Indeed, the condition that the r.h.s. of (9) also weakly vanish is easily verified to imply, upon expansion in the derivatives of the gauge parameter ϵ, that $\overline{\psi}_\mu\Gamma^{\mu\alpha\beta\gamma\delta}\epsilon F_{\alpha\beta\gamma\delta}$ must vanish on shell, which it does ***not*** do.

Can one improve the first-order deformation (8) to make it acceptable? The cosmological term will not help because it does not transform into F. The only possible candidates would be functions of the 3-form field. In order to define observables, these functions must be invariant under the gauge transformations of the 3-form, at least on-shell and up to a total derivative. However, in 11 dimensions, the only such functions can be redefined so as to

---

[d]We emphasize that in this procedure, one cannot start with the cosmological term as a $\Delta S^{(1)}$. Indeed, the variation of the cosmological term under the gauge transformations of the undeformed theory is algebraic in the fields and hence does not vanish on-shell, even up to a surface term. Hence it is not an observable of the undeformed theory, and so cannot be a starting point for a consistent deformation: adding the cosmological term (or the sum of it and the mass term) as a $\Delta S^{(1)}$ to the ordinary supergravity action is a much more radical (indeed inconsistent !) change than the gravitino mass term alone.
[e]Many of the features of (7) were anticipated in [24].

be off-shell (and not just on-shell) gauge invariant, up to a total derivative. This follows from an argument that closely patterns the analysis of [25], defining the very restricted class of on-shell invariant vertices that cannot in general be extended off-shell. [The above result actually justifies the non-trivial assumption of [10], that "on–" implies "off–".] Thus, the available functions of A may be assumed to be strictly gauge invariant, i.e., to be functions of the field strength F (which eliminates A^2; also, changing the coefficient of the Chern-Simons term in the original action clearly cannot help). But it is easy to see that no expression in F can cancel the unwanted term in (9), because of a mismatch in the number of derivatives. Hence, there is no way to improve the mass term to turn it into an observable in 11 dimensions. It is the A-field part of the supersymmetry variation of the gravitino that is responsible for the failure of the mass term to be an observable, just as it was also responsible for the difficulties described in the first, linearized, approach. Since the cohomology procedure saves us from also seeking modifications of the transformations rules, we can conclude that the introduction of a cosmological constant is obstructed already at the first step in $D = 11$ supergravity from the full theory end as well.

In our discussion, we have assumed (as in lower dimensions) both that the limit of a vanishing mass m is smooth[f] and that the field content remains unchanged in the cosmological variant. Any "no-go" result is of course no stronger than its assumptions, and ours are shared by the earlier treatments [3,10] that we surveyed. There is one (modest) loosening that can be shown not to work either, inspired by a recent reformulation [26] of the $D = 10$ cosmological model [9]. The idea is to add a deformation involving a nonpropagating field, here the 11-form $G_{11} \equiv dA_{10}$, through an addition $\Delta I \sim \int(dx)[G_{11} + b\bar{\psi}\Gamma^9\psi]^2$. The A_{10}-field equation states that the dual, $\epsilon^{11}[G_{11} + b\bar{\psi}\Gamma^9\psi]$ is a constant of integration, say m. The resulting supergravity field equations look like the "cosmological" desired ones. However, while this "dualization" works for lower dimensions, in $D = 11$ we are simply back to the original inconsistent model with supersymmetry still irremediably lost, as can be also discovered –without integrating out– in the deformation approach.

I will end this account with a rather different set of "uniqueness" questions about possible extensions that we are currently attempting to settle. Here the invariants in question are those that could, if they existed, constitute possible (infinite) counterterms in a perturbative loop expansion of the theory. It is of course well-known that all supergravities in D$\geq$4 are power counting

[f] This restriction is not necessarily stringent: in cosmological $D = 10$ supergravity [9], there is m^{-1} dependence in a field transformation rule, but that is an artefact removable by introducing a Stuckelberg compensator.

nonrenormalizable *a priori*, since the underlying Einstein models are, so the question is whether supersymmetry can save the day. But already for D=4, N=1 it was shown early on [27] that at three loops and higher, suitable supersymmetric invariants existed, and it would be very unlikely if their coefficients precisely vanished in (seemingly impossible to perform!) explicit calculations. Now strictly speaking, before talking about candidate terms, one must first exhibit a regularization scheme that preserves the supersymmetry, something notoriously difficult in odd dimensions (due to the Levi–Civita symbol, for example). While we cannot point to dimensional regularization as a generally legitimate scheme, it certainly is at low loop orders so we carry on within it there. We then seek terms that are a) supersymmetric, b) dimensionally correct in a loop expansion in the sole dimensional constant of the theory, the gravitational one. Recall that the Einstein term $\kappa^{-2}R$ in D=11 fixes the dimension of κ^2 to be L^9. The constant κ also appears in front of the form field's famous Chern–Simons term, $\kappa\epsilon FFA$ (here parity preserving!), as is clear by comparing its dimension with those of the kinetic term F^2.

Since the purely gravitational parts of local counterterms, being of the form R^n (possibly involving a necessarily even number of covariant derivatives) are even-dimensional, only odd powers of κ^2 and hence only *even* loops can contribute to a local integral over $(d^{11}x)$. This "counting" fact has long been known (*e.g.*, [28]) although strictly speaking there does exist a gravitational Chern–Simons term of the form $\int \epsilon^{1...11} R_1..R_5\omega$ that has odd dimensions. However, it has odd parity and so should not arise in this parity even model at the one loop level where it could (at most) appear. One need only worry about 2k-loop invariants, and then indeed only about the subset of invariants that fail to vanish on-shell;[g] those that do vanish there can always be absorbed by a harmless field-redefinition [30]. The simplest, two-loop, contribution would presumably begin as $\kappa^{+2}\int d^{11}x\,\Delta L_2$, with the leading gravitational parts $\Delta L_2 \sim R^{10}+..+(D^3R^2)^2+..(D^8R)^2$ in a very schematic notation; the (on-shell) R's are all Weyl tensors and D represents a covariant derivative. Likewise the F-field would enter through invariants of suitable powers of F and their derivatives, in addition to dimensionally relevant fermionic and mixed terms, such as $RRFF$, $\bar{\psi}\psi(RR+FF)$ at four-particle, 2-loop, order. After the rest of this report was written, there has just appeared [31] a calculation that points to the existence of such a two-loop non-vanishing local counter-term for D=11. Its algebraic structure is still quite murky as only the purely gravitational $O(R^4)$ contribution is presented, and that in rather

[g] We emphasize that the present analysis is a perturbative one within the D=11 SUGRA framework only. Thus, other possible "invariants" that arise *e.g.*, in M-theory or through dimensional reduction but do not have integer powers of κ, will not surface here [29].

implicit form; presumably that is related to the square of the Bel–Robinson tensor, as is the case for the counterterms in lower dimensions [27]. It may perhaps also involve the quartic Gauss–Bonnet invariant, which even in $D > 8$ is a total divergence to lowest order and so would not contribute there. We are currently attempting to verify and formulate the full super-invariant into which the particular R^4 term can be incorporated. This is a nontrivial task purely algebraically; not only must the R^4 be expressed in a "SUSY-useful" manner, but one must then identify the corresponding four-point contributions into which it transforms under SUSY rotation. This process will also yield insight into how one could find invariants more generally in this model, which lacks a sufficiently useful formalism to provide them; it will also require a lot of algebraic technology in identifying the form contributions in particular, and we hope to report on the results in due course. It would have been more exciting had no such invariant proved possible, so that D=11 supergravity could have been a nice "M-theory" all on its own, but the evidence of [31] is, alas, pretty convincing.

This work was supported by NSF grant PHY 93-15811

References

1. K. Bautier, S. Deser, M. Henneaux, and D. Seminara, Phys. Lett. **B406** (1997) 49.
2. E. Cremmer, B. Julia and J. Scherk, *Phys. Lett.* **B76**, (1978) 409. We use this paper's conventions.
3. W. Nahm, *Nucl. Phys.* **B135** (1978) 145 and references therein.
4. R. Troncoso and J. Zanelli, *Phys. Rev.* **D58** (1998) 101703.
5. C. Aragone and S. Deser, *Phys. Lett.* **86B** (1979) 161; *Nuov. Cim.* **57B** (1980) 33.
6. D. Boulware and S. Deser, *Phys. Rev.* **D6** (1972) 3368.
7. P. K. Townsend, *Phys. Rev.* **D15** (1977) 2802.
8. S. Deser and B. Zumino, *Phys. Rev. Lett.* **38** (1977) 1433.
9. L. Romans, *Phys. Lett.* **B169** (1986) 374.
10. A. Sagnotti and T. N. Tomaras, *Properties of 11-Dimensional Supergravity*, Caltech preprint CALT-68-885 (1982) unpublished.
11. R. D'Auria and P. Fre, *Nucl. Phys.* **B201** (1982) 101.
12. P.S. Howe, N.D. Lambert, and P.C. West, *Phys. Lett.* **B416**(1998) 303.
13. O. Bergmann, M.R. Gaberdiel and G. Lifschytz, *Nucl. Phys.* **B509** (1998) 194; C.M. Hull, *Nucl. Phys.* **B509** (1998) 216.
14. H. Nicolai, P.K. Townsend and P. van Nieuwenhuizen, *Lett. Nuov. Cim.*

30 (1981) 315 .
15. G. Barnich and M. Henneaux, *Phys.Lett.* **B311** (1993) 123.
16. The usefulness of the deformation point of view (but not in the general framework of the antifield formalism, which allows off-shell open deformations of the algebra) has been advocated in B. Julia, in *Recent Developments in Quantum Field Theory*, J. Ambjorn, B.J. Durhuus and J. L. Petersen eds, Elsevier (1985) pp 215-225; B. Julia, in *Topological and Geometrical Methods in Field Theory*, J. Hietarinta and J. Westerholm eds, World Scientific (1986) pp 325-339.
17. J. Stasheff, *Deformation Theory and the Batalin-Vilkovisky Master Equation*, q-alg/9702012.
18. M. Henneaux and C. Teitelboim, *Quantization of gauge systems*, (Princeton University Press, Princeton, 1992).
19. J. Zinn-Justin, *Renormalization of Gauge Theories*, in: Trends in Elementary Particle Theory, Lecture Notes in Physics 37, Springer, Berlin 1975; I. A. Batalin and G. A. Vilkovisky, *Phys. Lett.* **B102** (1981) 27; *Phys. Rev.* **D28** (1983) 2567; J. Gomis, J. Paris and S. Samuel, *Phys. Rep.* **259** (1995) 1.
20. J. Gomis and S. Weinberg, *Nucl. Phys.* **B469** (1996) 473.
21. S. Deser and B. Zumino, *Phys. Lett.* **B62** (1976) 335; D.Z. Freedman, P. van Nieuwenhuizen and S. Ferrara, *Phys. Rev.* **D13** (1976) 3214.
22. R. Kallosh, *Nucl. Phys.* **B141** (1978) 141.
23. F. Brandt, *Ann. Phys.* **259** (1997) 253.
24. B. de Wit, P. van Nieuwenhuizen and A. Van Proeyen, *Phys. Lett.* **B104** (1981) 27.
25. M. Henneaux, *Phys. Lett.* **B368** (1996) 83; M. Henneaux, B. Knaepen and C. Schomblond, *Commun. Math. Phys.* **186** (1997) 137.
26. E. Bergshoeff, M. de Roo, G. Papadopoulos, M.B. Green and P.K. Townsend, *Nucl. Phys.* **B470** (1996) 113.
27. S. Deser, J.H. Kay, and K.S. Stelle, *Phys. Rev. Lett.* **38** (1977) 527.
28. M.J. Duff and D.J. Toms in *Unification of the Fundamental Particle Interactions* (eds. Ellis and Ferrara, Plenum 1982).
29. M.J. Duff, J. T. Liu and R. Minasian, *Nucl. Phys.* **B452** (1995) 261; M.J. Duff, *Supermembranes*, hep-th/9611203; M. Green, M. Gutperle, and P. Vanhove, *Phys. Lett.* **B409** (1997) 177.
30. G. 't Hooft in *Acta Univ. Wratislavensis No. 368*, Proc. of XII Winter School, Karpacz.
31. Z. Bern, L. Dixon, D.C. Dunbar, M. Perelstein and J.S. Rozowsky, *Nucl. Phys.* **B530** (1998) 401.

PROBING BLACK HOLES AND RELATIVISTIC STARS WITH GRAVITATIONAL WAVES

KIP S. THORNE

California Institute of Technology, Pasadena, CA 91125 USA

In the coming decade, gravitational waves will convert the study of general relativistic aspects of black holes and stars from a largely theoretical enterprise to a highly interactive, observational/theoretical one. For example, gravitational-wave observations should enable us to observationally map the spacetime geometries around quiescent black holes, study quantitatively the highly nonlinear vibrations of curved spacetime in black-hole collisions, probe the structures of neutron stars and their equation of state, search for exotic types of general relativistic objects such as boson stars, soliton stars, and naked singularities, and probe aspects of general relativity that have never yet been seen such as the gravitational fields of gravitons and the influence of gravitational-wave tails on radiation reaction.

1 Introduction

In this article, I shall describe the prospects for using gravitational waves to probe the warpage of spacetime around black holes and relativistic stars, and to search for new types of general relativistic objects, for which there as yet is no observational evidence. And I shall describe how the challenge of developing data analysis algorithms for gravitational-wave detectors is already driving the theory of black holes and relativistic stars just as hard as the theory is driving the wave-detection efforts. Already, several years before the full-scale detectors go into operation, the challenge of transforming general relativistic astrophysics into an observational science has transformed the nature of our theoretical enterprise. At last, after 35 years of only weak coupling to observation, those of us studying general relativistic aspects of black holes and stars have become tightly coupled to the observational/experimental enterprise.

2 Gravitational Waves

A gravitational wave is a ripple of warpage (curvature) in the "fabric" of spacetime. According to general relativity, gravitational waves are produced by the dynamical spacetime warpage of distant astrophysical systems, and they travel outward from their sources and through the Universe at the speed of light, becoming very weak by the time they reach the Earth. Einstein discovered gravitational waves as a prediction of his general relativity theory in 1916, but only in the late 1950s did the technology of high-precision measurement become good enough to justify an effort to construct detectors for the waves.

Gravitational-wave detectors and detection techniques have now been under development for nearly 40 years, building on foundations laid by Joseph Weber [1], Rainer Weiss [2], and others. These efforts have led to promising sensitivities in four frequency bands, and theoretical studies have identified plausible sources in each band:

- The Extremely Low Frequency Band (ELF), 10^{-15} to 10^{-18} Hz, in which the measured anisotropy of the cosmic microwave background radiation places strong limits on gravitational wave strengths—and may, in fact, have detected waves [3,4]. The only waves expected in this band are relics of the big bang, a subject beyond the scope of this article. (For some details and references see [3,4,5] and references cited therein.)

- The Very Low Frequency Band (VLF), 10^{-7} to 10^{-9} Hz, in which Joseph Taylor and others have achieved remarkable gravity-wave sensitivities by the timing of millisecond pulsars [6]. The only expected strong sources in this band are processes in the very early universe—the big bang, phase transitions of the vacuum states of quantum fields, and vibrating or colliding defects in the structure of spacetime, such as monopoles, cosmic strings, domain walls, textures, and combinations thereof [7,8,9,10]. These sources are also beyond the scope of this article.

- The Low-Frequency Band (LF), 10^{-4} to 1 Hz, in which will operate the Laser Interferometer Space Antenna, LISA; see Sec. 3.4 below. This is the band of massive black holes ($M \sim 1000$ to $10^8 M_\odot$) in the distant universe, and of other hypothetical massive exotic objects (naked singularities, soliton stars), as well as of binary stars (ordinary, white dwarf, neutron star, and black hole) in our galaxy. Early universe processes should also have produced waves at these frequencies, as in the ELF and VLF bands.

- The High-Frequency Band (HF), 1 to 10^4Hz, in which operate earth-based gravitational-wave detectors such as LIGO; see Secs. 3.1–3.3 below. This is the band of stellar-mass black holes ($M \sim 1$ to $1000 M_\odot$) and of other conceivable stellar-mass exotic objects (naked singularities and boson stars) in the distant universe, as well as of supernovae, pulsars, and coalescing and colliding neutron stars. Early universe processes should also have produced waves at these frequencies, as in the ELF, VLF, and LF bands.

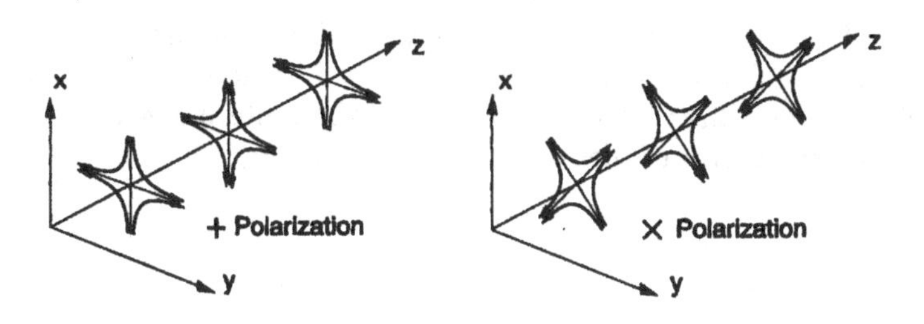

Figure 1. The lines of force associated with the two polarizations of a gravitational wave. (From Ref. [12].)

In this article I shall focus on the HF and LF bands, because these are the ones in which we can expect to study black holes and relativistic stars.

One aspect of a gravitational wave's spacetime warpage—the only aspect relevant to earth-based detectors—is an oscillatory "stretching and squeezing" of space. This stretch and squeeze is described, in general relativity theory, by two dimensionless gravitational wave fields h_+ and $h_\times$ (the "strains of space") that are associated with the wave's two linear polarizations, conventionally called "plus" (+) and "cross" (×). The fields h_+ and $h_\times$, technically speaking, are the double time integrals of space-time-space-time components of the Riemann curvature tensor; and they propagate through spacetime at the speed of light. The inertia of any small piece of an object tries to keep it at rest in, or moving at constant speed through, the piece of space in which it resides; so as h_+ and $h_\times$ stretch and squeeze space, inertia stretches and squeezes objects that reside in that space. This stretch and squeeze is analogous to the tidal gravitational stretch and squeeze exerted on the Earth by the Moon, and thus the associated gravitational-wave force is referred to as a "tidal" force.

If an object is small compared to the waves' wavelength (as is the case for ground-based detectors), then relative to the object's center, the waves exert tidal forces with the quadrupolar patterns shown in Fig. 1. The names "plus" and "cross" are derived from the orientations of the axes that characterize the force patterns [11].

The strengths of the waves from a gravitational-wave source can be estimated using the "Newtonian/quadrupole" approximation to the Einstein field equations. This approximation says that $h \simeq (G/c^4)\ddot{Q}/r$, where $\ddot{Q}$ is the second time derivative of the source's quadrupole moment, r is the distance of the source from Earth (and G and c are Newton's gravitation constant and the speed of light). The strongest sources will be highly nonspherical and thus will have $Q \simeq ML^2$, where M is their mass and L their size, and correspondingly will have $\ddot{Q} \simeq 2Mv^2 \simeq 4E_{\rm kin}^{\rm ns}$, where v is their internal velocity and $E_{\rm kin}^{\rm ns}$ is the nonspherical part of their internal kinetic energy. This provides us with the estimate

$$h \sim \frac{1}{c^2} \frac{4G(E_{\rm kin}^{\rm ns}/c^2)}{r} ; \tag{1}$$

i.e., h is about 4 times the gravitational potential produced at Earth by the mass-equivalent of the source's nonspherical, internal kinetic energy—made dimensionless by dividing by c^2. Thus, in order to radiate strongly, the source must have a very large, nonspherical, internal kinetic energy.

The best known way to achieve a huge internal kinetic energy is via gravity; and by energy conservation (or the virial theorem), any gravitationally-induced kinetic energy must be of order the source's gravitational potential energy. A huge potential energy, in turn, requires that the source be very compact, not much larger than its own gravitational radius. Thus, the strongest gravity-wave sources must be highly compact, dynamical concentrations of large amounts of mass (e.g., colliding and coalescing black holes and neutron stars).

Such sources cannot remain highly dynamical for long; their motions will be stopped by energy loss to gravitational waves and/or the formation of an all-encompassing black hole. Thus, the strongest sources should be transient. Moreover, they should be very rare—so rare that to see a reasonable event rate will require reaching out through a substantial fraction of the Universe. Thus, just as the strongest radio waves arriving at Earth tend to be extragalactic, so also the strongest gravitational waves are likely to be extragalactic.

For highly compact, dynamical objects that radiate in the high-frequency band, e.g. colliding and coalescing neutron stars and stellar-mass black holes, the internal, nonspherical kinetic energy $E_{\rm kin}^{\rm ns}/c^2$ is of order the mass of the Sun; and, correspondingly, Eq. (1) gives $h \sim 10^{-22}$ for such sources at the Hubble distance (3000 Mpc, i.e., 10^{10} light years); $h \sim 10^{-21}$ at 200 Mpc (a best-guess distance for several neutron-star coalescences per year; see Section 6.2), $h \sim 10^{-20}$ at the Virgo cluster of galaxies (15 Mpc); and $h \sim 10^{-17}$ in the outer reaches of our own Milky Way galaxy (20 kpc). These numbers set the scale of sensitivities that ground-based interferometers seek to achieve: $h \sim 10^{-21}$ to 10^{-22}.

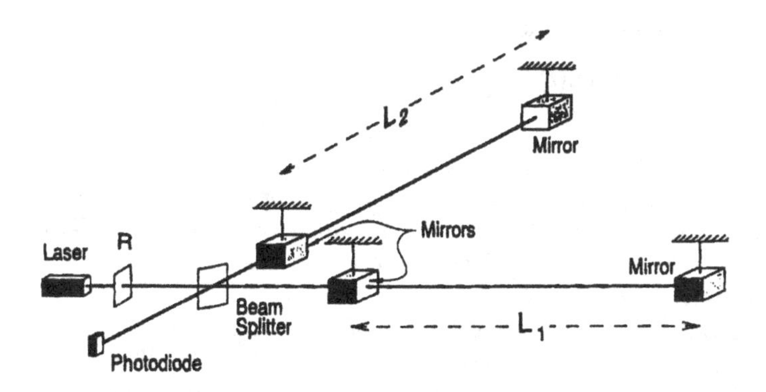

Figure 2. Schematic diagram of a laser interferometer gravitational wave detector. (From Ref. [12].)

3 Gravitational Wave Detectors in the High and Low Frequency Bands

3.1 Ground-Based Laser Interferometers

The most promising and versatile type of gravitational-wave detector in the high-frequency band, 1 to 10^4Hz, is a laser interferometer gravitational wave detector ("interferometer" for short). Such an interferometer consists of four mirror-endowed masses that hang from vibration-isolated supports as shown in Fig. 2, and the indicated optical system for monitoring the separations between the masses [11,12]. Two masses are near each other, at the corner of an "L", and one mass is at the end of each of the L's long arms. The arm lengths are nearly equal, $L_1 \simeq L_2 = L$. When a gravitational wave, with frequencies high compared to the masses' $\sim$ 1 Hz pendulum frequency, passes through the detector, it pushes the masses back and forth relative to each other as though they were free from their suspension wires, thereby changing the arm-length difference, $\Delta L \equiv L_1 - L_2$. That change is monitored by laser interferometry in such a way that the variations in the output of the photodiode (the interferometer's output) are directly proportional to $\Delta L(t)$.

If the waves are coming from overhead or underfoot and the axes of the + polarization coincide with the arms' directions, then it is the waves' + polarization that drives the masses, and the detector's strain $\Delta L(t)/L$ is equal to the waves' strain of space $h_+(t)$. More generally, the interferometer's output is a linear combination of the two wave fields:

$$\frac{\Delta L(t)}{L} = F_+ h_+(t) + F_\times h_\times(t) \equiv h(t) \ . \qquad (2)$$

The coefficients F_+ and $F_\times$ are of order unity and depend in a quadrupolar manner on the direction to the source and the orientation of the detector [11]. The combination $h(t)$ of the two h's is called the gravitational-wave strain that acts on the detector; and the time evolutions of $h(t)$, $h_+(t)$, and $h_\times(t)$ are sometimes called *waveforms*.

When one examines the technology of laser interferometry, one sees good prospects to achieve measurement accuracies $\Delta L \sim 10^{-16}$ cm (1/1000 the diameter of the nucleus of an atom)—and $\Delta L = 8 \times 10^{-16}$ has actually been achieved in a prototype interferometer at Caltech [13]. With $\Delta L \sim 10^{-16}$cm, an interferometer must have an arm length $L = \Delta L/h \sim 1$ to 10 km in order to achieve the desired wave sensitivities, 10^{-21} to 10^{-22}. This sets the scale of the interferometers that are now under construction.

3.2 LIGO, VIRGO, and the International Network of Gravitational Wave Detectors

Interferometers are plagued by non-Gaussian noise, e.g. due to sudden strain releases in the wires that suspend the masses. This noise prevents a single interferometer, by itself, from detecting with confidence short-duration gravitational-wave bursts (though it may be possible for a single interferometer to search for the periodic waves from known pulsars). The non-Gaussian noise can be removed by cross correlating two, or preferably three or more, interferometers that are networked together at widely separated sites.

The technology and techniques for such interferometers have been under development for 25 years, and plans for km-scale interferometers have been developed over the past 15 years. An international network consisting of three km-scale interferometers at three widely separated sites is now under construction. It includes two sites of the American LIGO Project ("Laser Interferometer Gravitational Wave Observatory") [12], and one site of the French/Italian VIRGO Project (named after the Virgo cluster of galaxies) [14].

LIGO will consist of two vacuum facilities with 4-kilometer-long arms, one in Hanford, Washington (in the northwestern United States) and the other in Livingston, Louisiana (in the southeastern United States). These facilities are designed to house many successive generations of interferometers without the necessity of any major facilities upgrade; and after a planned future expansion, they will be able to house several interferometers at once, each with a different optical configuration optimized for a different type of wave (e.g., broad-band burst, or narrow-band periodic wave, or stochastic wave).

The LIGO facilities are being constructed by a team of about 80 physicists and engineers at Caltech and MIT, led by Barry Barish (the PI), Gary Sanders (the Project Manager), Albert Lazzarini, Rai Weiss, Stan Whitcomb, and Robbie Vogt (who directed the project during the pre-construction phase). This Caltech/MIT team, together with researchers from several other universities, is developing LIGO's first interferometers and their data analysis system. Other research groups from many universities are contributing to R&D for *enhancements* of the first interferometers, or are computing theoretical waveforms for use in data analysis, or are developing data analysis techniques for future interferometers. These groups are linked together in a *LIGO Scientific Collaboration* and by an organization called the *LIGO Research Community*. For further details, see the LIGO World Wide Web Site, http://www.ligo.caltech.edu/.

The VIRGO Project is building one vacuum facility in Pisa, Italy, with 3-kilometer-long arms. This facility and its first interferometers are a collaboration of more than a hundred physicists and engineers at the INFN (Frascati, Napoli, Perugia, Pisa), LAL (Orsay), LAPP (Annecy), LOA (Palaiseau), IPN (Lyon), ESPCI (Paris), and the University of Illinois (Urbana), under the leadership of Alain Brillet and Adalberto Giazotto.

The LIGO and VIRGO facilities are scheduled for completion at the end of the 1990's, and their first gravitational-wave searches will be performed in 2001 or 2002. Figure 3 shows the design sensitivities for LIGO's *first interferometers* (ca. 2001) [12] and for *enhanced versions* of those interferometers (which are expected to be operating five years or so later) [15], along with a benchmark sensitivity goal for subsequent, more *advanced interferometers* [12,15].

For each type of interferometer, the quantity shown is the "sensitivity to bursts" that come from a random direction, $h_{\rm SB}(f)$ [12]. This $h_{\rm SB}$ is about 5 times worse than the rms noise level in a bandwidth $\Delta f \simeq f$ for waves with a random direction and polarization, and about $5\sqrt{5} \simeq 11$ times worse than the the rms noise level $h_{\rm rms}$ for optimally directed and polarized waves. (In much of the literature, the quantity plotted is $h_{\rm rms} \simeq h_{\rm SB}/11$.) Along the right-hand branch of each sensitivity curve (above 100 or 200 Hz), the interferometer's dominant noise is due to photon counting statistics ("shot noise"); along the middle branch (10 or 30 Hz to 100 to 200 Hz), the dominant noise is random fluctuations of thermal energy in the test masses and their suspensions; along the steep left-hand branch, the dominant noise is seismic vibrations creeping through the interferometers' seismic isolation system.

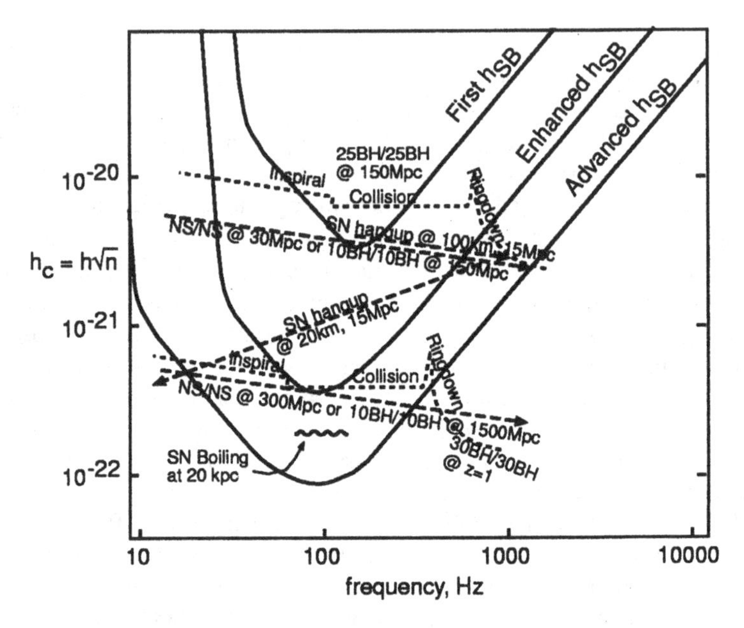

Figure 3. LIGO's projected broad-band noise sensitivity to bursts $h_{\rm SB}$ (Refs. [12,15]) compared with the characteristic amplitudes h_c of the waves from several hypothesized sources. The signal to noise ratios are $\sqrt{2}$ higher than in Ref. [12] because of a factor 2 error in Eq. (29) of Ref. [11].

The interferometer sensitivity $h_{\rm SB}$ is to be compared with the "characteristic amplitude" $h_c(f) = h\sqrt{n}$ of the waves from a source; here h is the waves' amplitude when they have frequency f, and n is the number of cycles the waves spend in a bandwidth $\Delta f \simeq f$ near frequency f [11,12]. Any source with $h_c > h_{\rm SB}$ should be detectable with high confidence, even if it arrives only once per year.

Figure 3 shows the estimated or computed characteristic amplitudes h_c for several sources that will be discussed in detail later in this article. Among these sources are binary systems made of $1.4M_\odot$ neutron stars ("NS") and binaries made of 10, 25, and 30 $M_\odot$ black holes ("BH"), which spiral together and collide under the driving force of gravitational radiation reaction. As the bodies spiral inward, their waves sweep upward in frequency (rightward across the figure along the dashed lines). From the figure we see that LIGO's first interferometers should be able to detect waves from the inspiral of a NS/NS binary out to a distance of 30Mpc (90 million light years) and from the final collision and merger of a $25M_\odot/25M_\odot$ BH/BH binary out to about 300Mpc. Comparison with estimated event rates (Secs. 6.2 and 7.2 below) suggests, with considerable confidence, that the first wave detections will be achieved by the time the enhanced sensitivity is reached and possibly as soon as the first-interferometers' searches.

LIGO alone, with its two sites which have parallel arms, will be able to detect an incoming gravitational wave, measure one of its two waveforms, and (from the time delay between the two sites) locate its source to within a $\sim 1^\circ$ wide annulus on the sky. LIGO and VIRGO together, operating as a ***coordinated international network***, will be able to locate the source (via time delays plus the interferometers' beam patterns) to within a 2-dimensional error box with size between several tens of arcminutes and several degrees, depending on the source direction and on the amount of high-frequency structure in the waveforms. They will also be able to monitor both waveforms $h_+(t)$ and $h_\times(t)$ (except for frequency components above about 1kHz and below about 10 Hz, where the interferometers' noise becomes severe).

A British/German group is constructing a 600-meter interferometer called GEO 600 near Hanover Germany [16], and Japanese groups, a 300-meter interferometer called TAMA near Tokyo [17]. GEO600 may be a significant player in the interferometric network in its early years (by virtue of cleverness and speed of construction), but because of its short arms it cannot compete in the long run. GEO600 and TAMA will both be important development centers and testbeds for interferometer techniques and technology, and in due course they may give rise to kilometer-scale interferometers like LIGO and VIRGO, which could significantly enhance the network's all-sky coverage and ability to extract information from the waves.

3.3 Narrow-Band, High-Frequency Detectors: Interferometers and Resonant-Mass Antennas

At frequencies $f \gtrsim 500$Hz, the interferometers' photon shot noise becomes a serious obstacle to wave detection. However, narrow-band detectors specially optimized for kHz frequencies show considerable promise. These include interferometers with specialized optical configurations ("signal recycled interferometers" [18] and "resonant sideband extraction interferometers" [19]), and large spherical or truncated icosahedral resonant-mass detectors (e.g., the American TIGA [20], Dutch GRAIL [21] and Brazilian OMNI-1 Projects) that are future variants of Joseph Weber's original "bar" detector [1] and of currently operating bars in Italy (AURIGA, Explorer and Nautilus), Australia (NIOBE), and America (ALLEGRO) [22]. Developmental work for these narrow-band

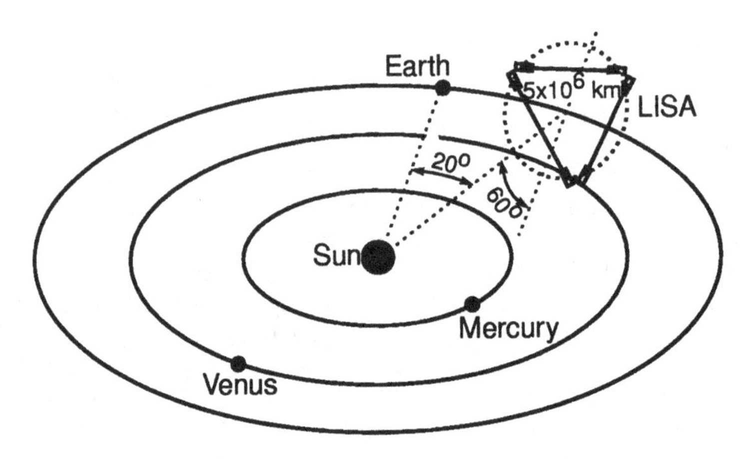

Figure 4. LISA's orbital configuration, with LISA magnified in arm length by a factor ~ 10 relative to the solar system.

detectors is underway at a number of centers around the world.

3.4 Low-Frequency Detectors—The Laser Interferometer Space Antenna (LISA)

The *Laser Interferometer Space Antenna* (LISA) [23] is the most promising detector for gravitational waves in the low-frequency band, 10^{-4}–1 Hz (10,000 times lower than the LIGO/VIRGO high-frequency band).

LISA was originally conceived (under a different name) by Peter Bender of the University of Colorado, and is currently being developed by an international team led by Karsten Danzmann of the University of Hanover (Germany) and James Hough of Glasgow University (UK). The European Space Agency tentatively plans to fly it sometime in the 2014–2018 time frame as part of ESA's Horizon 2000+ Program of large space missions. With NASA participation (which is under study), the flight could be much sooner.

As presently conceived [23], LISA will consist of six compact, drag-free spacecraft (i.e. spacecraft that are shielded from buffeting by solar wind and radiation pressure, and that thus move very nearly on geodesics of spacetime). All six spacecraft would be launched simultaneously in a single Ariane rocket. They would be placed into the same heliocentric orbit as the Earth occupies, but would follow 20° behind the Earth; cf. Fig. 4. The spacecraft would fly in pairs, with each pair at the vertex of an equilateral triangle that is inclined at an angle of 60° to the Earth's orbital plane. The triangle's arm

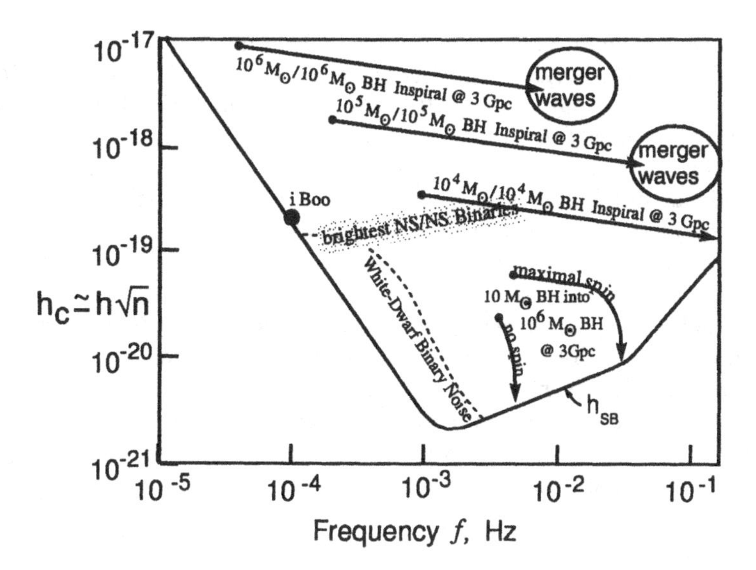

Figure 5. LISA's projected sensitivity to bursts h_{SB}, compared with the strengths of the waves from several low-frequency sources [23].

length would be 5 million km (10^6 times longer than LIGO's arms!). The six spacecraft would track each other optically, using one-Watt YAG laser beams. Because of diffraction losses over the 5×10^6km arm length, it is not feasible to reflect the beams back and forth between mirrors as is done with LIGO. Instead, each spacecraft would have its own laser; and the lasers would be phase locked to each other, thereby achieving the same kind of phase-coherent out-and-back light travel as LIGO achieves with mirrors. The six-laser, six-spacecraft configuration thereby would function as three, partially independent and partially redundant, gravitational-wave interferometers.

Figure 5 depicts the expected sensitivity of LISA in the same language as we have used for LIGO (Fig. 3): $h_{SB} = 5\sqrt{5}h_{rms}$ is the sensitivity for high-confidence detection ($S/N = 5$) of a signal coming from a random direction, assuming Gaussian noise.

At frequencies $f \gtrsim 10^{-3}$Hz, LISA's noise is due to photon counting statistics (shot noise). The sensitivity curve steepens at $f \sim 3 \times 10^{-2}$Hz because at larger f than that, the waves' period is shorter than the round-trip light travel time in one of LISA's arms. Below 10^{-3}Hz, the noise is due to buffeting-induced random motions of the spacecraft that are not being properly removed by the drag-compensation system. Notice that, in terms of dimensionless amplitude, LISA's sensitivity is roughly the same as that of LIGO's first interferometers (Fig. 3), but at 100,000 times lower frequency. Since the waves' energy flux scales as f^2h^2, this corresponds to 10^{10} better energy sensitivity than LIGO.

LISA can detect and study, simultaneously, a wide variety of different sources scattered over all directions on the sky. The key to distinguishing the different sources is the different time evolution of their waveforms. The key to determining each source's direction, and confirming that it is real and not just noise, is the manner in which its waves' amplitude and frequency are modulated by LISA's complicated orbital motion—a motion in which the interferometer triangle rotates around its center once per year, and the interferometer plane precesses around the normal to the Earth's orbit once per year. Most sources will be observed for a year or longer, thereby making full use of these modulations.

4 Stellar Core Collapse: The Births of Neutron Stars and Black Holes

In the remainder of this article, I shall describe the techniques and prospects for observationally studying black holes and relativistic stars via the gravitational waves they emit. I begin with the births of stellar-mass neutron stars and black holes.

When the core of a massive star has exhausted its supply of nuclear fuel, it collapses to form a neutron star or a black hole. In some cases, the collapse triggers and powers a subsequent explosion of the star's mantle—a supernova explosion. Despite extensive theoretical efforts for more than 30 years, and despite wonderful observational data from Supernova 1987A, theorists are still far from a definitive understanding of the details of the collapse and explosion. The details are highly complex and may differ greatly from one core collapse to another [24].

Several features of the collapse and the core's subsequent evolution can produce significant gravitational radiation in the high-frequency band. We shall consider these features in turn, the most weakly radiating first, and we shall focus primarily on collapses that produce neutron stars rather than black holes.

4.1 Boiling of a Newborn Neutron Star

Even if the collapse is spherical, so it cannot radiate any gravitational waves at all, it should produce a convectively unstable neutron star that "boils" vigorously (and nonspherically) for the first ~ 1 second of its life [25]. The boiling dredges up high-temperature nuclear matter ($T \sim 10^{12}$K) from the neutron star's central regions, bringing it to the surface (to the "neutrino-sphere"), where it cools by neutrino emission before being swept back downward and reheated. Burrows [26] has pointed out that the boiling should generate $n \sim 100$ cycles of gravitational waves with frequency $f \sim 100$Hz and amplitude large enough to be detectable by LIGO/VIRGO throughout our galaxy and its satellites. Neutrino detectors have a similar range, and there could be a high scientific payoff from correlated observations of the gravitational waves emitted by the boiling's mass motions and neutrinos emitted from the boiling neutrino-sphere. With neutrinos to trigger on, the sensitivities of LIGO detectors should be about twice as good as shown in Fig. 3.

Recent 3+1 dimensional simulations by Müller and Janka [27] suggest an rms amplitude $h \sim 2 \times 10^{-23}(20\text{kpc}/r)$ (where r is the distance to the source), corresponding to a characteristic amplitude $h_c \simeq h\sqrt{n} \sim 2 \times 10^{-22}(20\text{kpc}/r)$; cf. Fig. 3. (The older 2+1 dimensional simulations gave h_c about 6 times larger than this [27], but presumably were less reliable.) LIGO should be able to detect such waves throughout our galaxy with an amplitude signal to noise ratio of about $S/N = 2.5$ in each of its two enhanced 4km interferometers, and its advanced interferometers should do the same out to 80Mpc distance. (Recall that the h_{SB} curves in Fig. 3 are drawn at a signal to noise ratio of about 5). Although the estimated event rate is only about one every 40 years in our galaxy and not much larger out to 80Mpc, if just one such supernova is detected the correlated neutrino and gravitational wave observations could bring very interesting insights into the boiling of a newborn neutron star.

4.2 Axisymmetric Collapse, Bounce, and Oscillations

Rotation will centrifugally flatten the collapsing core, enabling it to radiate as it implodes. If the core's angular momentum is small enough that centrifugal forces do not halt or strongly slow the collapse before it reaches nuclear densities, then the core's collapse, bounce, and subsequent oscillations are likely to be axially symmetric. Numerical simulations [28,29] show that in this case the waves from collapse, bounce, and oscillation will be quite weak: the total energy radiated as gravitational waves is not likely to exceed $\sim 10^{-7}$ solar masses (about 1 part in a million of the collapse energy) and might often be much less than this; and correspondingly, the waves' characteristic amplitude will be $h_c \lesssim 3 \times 10^{-21}(30\text{kpc}/r)$. These collapse-and-bounce waves will come off at frequencies $\sim$ 200 Hz to $\sim$ 1000 Hz, and will precede the boiling waves by a fraction of a second. Though a little stronger than the boiling waves, they probably cannot be seen by LIGO/VIRGO beyond the local group of galaxies and thus will be a very rare occurrence.

4.3 Rotation-Induced Bars and Break-Up

If the core's rotation is large enough to strongly flatten the core before or as it reaches nuclear density, then a dynamical or secular instability is likely to break the core's axisymmetry. The core will be transformed into a bar-like configuration that spins end-over-end like an American football, and that might even break up into two or more massive pieces. As we shall see below, the radiation from the spinning bar or orbiting pieces *could* be almost as strong as that from a coalescing neutron-star binary (Sec. 6.2), and thus could be seen by the LIGO/VIRGO first interferometers out to the distance of the Virgo cluster (where the supernova rate is several per year), by enhanced interferometers out to $\sim$ 100Mpc (supernova rate several thousand per year), and by advanced interferometers out to several hundred Mpc (supernova rate $\sim$ (a few) $\times\, 10^4$ per year); cf. Fig. 3. It is far from clear what fraction of collapsing cores will have enough angular momentum to break their axisymmetry, and what fraction of those will actually radiate at this high rate; but even if only $\sim 1/1000$ or $1/10^4$ do so, this could ultimately be a very interesting source for LIGO/VIRGO.

Several specific scenarios for such non-axisymmetry have been identified:

Centrifugal hangup at $\sim$ 100km radius: If the pre-collapse core is rapidly spinning (e.g., if it is a white dwarf that has been spun up by accretion from a companion), then the collapse may produce a highly flattened, centrifugally supported disk with most of its mass at radii $R \sim 100$km, which then (via instability) may transform itself into a bar or may bifurcate. The bar or bifurcated lumps will radiate gravitational waves at twice their rotation frequency, $f \sim 100$Hz—the optimal frequency for LIGO/VIRGO interferometers. To shrink on down to $\sim$ 10km size, this configuration must shed most of its angular momentum. *If* a substantial fraction of the angular momentum goes into gravitational waves, then independently of the strength of the bar, the waves will be nearly as strong as those from a coalescing binary. The reason is this: The waves' amplitude h is proportional to the bar's ellipticity e, the number of cycles n of wave emission is proportional to $1/e^2$, and the characteristic amplitude $h_c = h\sqrt{n}$ is thus independent of the ellipticity and is about the same whether the configuration is a bar or is two lumps [30]. The resulting waves will thus have h_c roughly half as large, at $f \sim 100$Hz, as the h_c from a NS/NS binary (half as large because each lump might be half as massive as a NS), and the waves will chirp upward in frequency in a manner similar to those from a binary (Sec. 6.2).

It may very well be, however, that most of the core's excess angular momentum does *not* go into gravitational waves, but instead goes largely into hydrodynamic waves as the bar or lumps, acting like a propeller, stir up the surrounding stellar mantle. In this case, the radiation will be correspondingly weaker.

Centrifugal hangup at $\sim$ 20km radius: Lai and Shapiro [31] have explored the case of centrifugal hangup at radii not much larger than the final neutron star, say $R \sim 20$km. Using compressible ellipsoidal models, they have deduced that, after a brief period of dynamical bar-mode instability with wave emission at $f \sim 1000$Hz (explored by Houser, Centrella, and Smith [32]), the star switches to a secular instability in which the bar's angular velocity gradually slows while the material of which it is made retains its high rotation speed and circulates through the slowing bar. The slowing bar emits waves that sweep *downward* in frequency through the LIGO/VIRGO optimal band $f \sim 100$Hz, toward ~ 10Hz. The characteristic amplitude (Fig. 3) is only modestly smaller than for the upward-sweeping waves from hangup at $R \sim 100$km, and thus such waves should be detectable near the Virgo Cluster by the first LIGO/VIRGO interferometers, near 100Mpc by enhanced interferometers, and at distances of a few 100Mpc by advanced interferometers.

Successive fragmentations of an accreting, newborn neutron star: Bonnell and Pringle [33] have focused on the evolution of the rapidly spinning, newborn neutron star as it quickly accretes more and more mass from the pre-supernova star's inner mantle. If the accreting material carries high angular momentum, it may trigger a renewed bar formation, lump formation, wave emission, and coalescence, followed by more accretion, bar and lump formation, wave emission, and coalescence. Bonnell and Pringle speculate that hydrodynamics, not wave emission, will drive this evolution, but that the total energy going into gravitational waves might be as large as $\sim 10^{-3} M_\odot$. This corresponds to $h_c \sim 10^{-21}(10\text{Mpc}/r)$.

5 Pulsars: Spinning Neutron Stars

As the neutron star settles down into its final state, its crust begins to solidify (crystalize). The solid crust will assume nearly the oblate axisymmetric shape that centrifugal forces are trying to maintain, with poloidal ellipticity $\epsilon_p \propto$(angular velocity of rotation)^2. However, the principal axis of the star's moment of inertia tensor may deviate from its spin axis by some small "wobble angle" θ_w, and the star may deviate slightly from axisymmetry about its principal axis; i.e., it may have a slight ellipticity $\epsilon_e \ll \epsilon_p$ in its equatorial plane.

As this slightly imperfect crust spins, it will radiate gravitational waves [34]: ϵ_e radiates at twice the rotation frequency, $f = 2f_{\rm rot}$ with $h \propto \epsilon_e$, and the wobble angle couples to ϵ_p to produce waves at $f = f_{\rm rot} + f_{\rm prec}$ (the precessional sideband of the rotation frequency) with amplitude $h \propto \theta_w \epsilon_p$. For typical neutron-star masses and moments of inertia, the wave amplitudes are

$$h \sim 6 \times 10^{-25} \left(\frac{f_{\rm rot}}{500{\rm Hz}}\right)^2 \left(\frac{1{\rm kpc}}{r}\right) \left(\frac{\epsilon_e \text{ or } \theta_w \epsilon_p}{10^{-6}}\right) . \tag{3}$$

The neutron star gradually spins down, due in part to gravitational-wave emission but perhaps more strongly due to electromagnetic torques associated with its spinning magnetic field and pulsar emission. This spin-down reduces the strength of centrifugal forces, and thereby causes the star's poloidal ellipticity ϵ_p to decrease, with an accompanying breakage and resolidification of its crust's crystal structure (a "starquake") [35]. In each starquake, θ_w, ϵ_e, and ϵ_p will all change suddenly, thereby changing the amplitudes and frequencies of the star's two gravitational "spectral lines" $f = 2f_{\rm rot}$ and $f = f_{\rm rot} + f_{\rm prec}$. After each quake, there should be a healing period in which the star's fluid core and solid crust, now rotating at different speeds, gradually regain synchronism. By monitoring the amplitudes, frequencies, and phases of the two gravitational-wave spectral lines, and by comparing with timing of the electromagnetic pulsar emission, one might learn much about the physics of the neutron-star interior.

How large will be the quantities ϵ_e and $\theta_w \epsilon_p$? Rough estimates of the crustal shear moduli and breaking strengths suggest an upper limit in the range $\epsilon_{\rm max} \sim 10^{-4}$ to 10^{-6}, and it might be that typical values are far below this. We are extremely ignorant, and correspondingly there is much to be learned from searches for gravitational waves from spinning neutron stars.

One can estimate the sensitivity of LIGO/VIRGO (or any other broadband detector) to the periodic waves from such a source by multiplying the waves' amplitude h by the square root of the number of cycles over which one might integrate to find the signal, $n = f\hat{\tau}$ where $\hat{\tau}$ is the integration time. The resulting effective signal strength, $h\sqrt{n}$, is larger than h by

$$\sqrt{n} = \sqrt{f\hat{\tau}} = 10^5 \left(\frac{f}{1000{\rm Hz}}\right)^{1/2} \left(\frac{\hat{\tau}}{4{\rm months}}\right)^{1/2} . \tag{4}$$

Four months of integration is not unreasonable in targeted searches; but for an all-sky, all-frequency search, a coherent integration might not last longer than a few days because of computational limitations associated with having to apply huge numbers of trial neutron-star spindown corrections and earth-motion doppler corrections [36].

Equations (3) and (4) for $h\sqrt{n}$ should be compared (i) to the detector's rms broad-band noise level for sources in a random direction, $\sqrt{5}h_{\rm rms}$, to deduce a signal-to-noise ratio, or (ii) to $h_{\rm SB}$ to deduce a sensitivity for high-confidence detection when one does not know the waves' frequency in advance [11]. Such a comparison suggests that the first interferometers in LIGO/VIRGO might possibly see waves from nearby spinning neutron stars, but the odds of success are very unclear.

The deepest searches for these nearly periodic waves will be performed by narrow-band detectors, whose sensitivities are enhanced near some chosen frequency at the price of sensitivity loss elsewhere—signal-recycled interferometers [18], resonant-sideband-extraction interferometers [19], or resonant-mass antennas [20,21] (Section 3.3). With "advanced-detector technology" and targeted searches, such detectors might be able to find with confidence spinning neutron stars that have [11]

$$(\epsilon_e \text{ or } \theta_w\epsilon_p) \gtrsim 3\times10^{-10}\left(\frac{500\text{Hz}}{f_{\rm rot}}\right)^2\left(\frac{r}{1000\text{pc}}\right)^2 . \qquad (5)$$

There may well be a large number of such neutron stars in our galaxy; but it is also conceivable that there are none. We are extremely ignorant.

Some cause for optimism arises from several physical mechanisms that might generate radiating ellipticities large compared to 3×10^{-10}:

- It may be that, inside the superconducting cores of many neutron stars, there are trapped magnetic fields with mean strength $B_{\rm core}\sim10^{13}$G or even 10^{15}G. Because such a field is actually concentrated in flux tubes with $B = B_{\rm crit}\sim6\times10^{14}$G surrounded by field-free superconductor, its mean pressure is $p_B = B_{\rm core}B_{\rm crit}/8\pi$. This pressure could produce a radiating ellipticity $\epsilon_e\sim\theta_w\epsilon_p\sim p_B/p\sim10^{-8}B_{\rm core}/10^{13}$G (where p is the core's material pressure).

- Accretion onto a spinning neutron star can drive precession (keeping θ_w substantially nonzero), and thereby might produce measurably strong waves [37].

- If a neutron star is born rotating very rapidly, then it may experience a gravitational-radiation-reaction-driven instability first discovered by Chandrasekhar [38] and elucidated in greater detail by Friedman and Schutz [39]). In this "CFS instability", density waves travel around the star in the opposite direction to its rotation, but are dragged forward by the rotation. These density waves produce gravitational waves that carry positive energy as seen by observers far from the star, but negative energy from the star's viewpoint; and because the star thinks it is losing negative energy, its density waves get amplified. This intriguing mechanism is similar to that by which spiral density waves are produced in galaxies. Although the CFS instability was once thought ubiquitous for spinning stars [39,40], we now know that neutron-star viscosity will kill it, stabilizing the star and turning off the waves, when the star's temperature is above some limit $\sim 10^{10}$K [41] and below some limit $\sim 10^9$K [42]; and correspondingly, the instability should operate only during the first few years of a neutron star's life, when $10^9\text{K} \lesssim T \lesssim 10^{10}\text{K}$.

6 Neutron-Star Binaries and Their Coalescence

6.1 NS/NS and Other Compact Binaries in Our Galaxy

The best understood of all gravitational-wave sources are binaries made of two neutron stars ("NS/NS binaries"). The famous Hulse-Taylor [43,44] binary pulsar, PSR 1913+16, is an example. At present PSR 1913+16 has an orbital frequency of about 1/(8 hours) and emits its waves predominantly at twice this frequency, roughly 10^{-4} Hz, which is in LISA's low-frequency band (Fig. 5); but it is too weak for LISA to detect. LISA will be able to search for brighter NS/NS binaries in our galaxy with periods shorter than this.

If conservative estimates [45,46,47] based on the statistics of binary pulsar observations are correct, there should be many NS/NS binaries in our galaxy that are brighter in gravitational waves than PSR 1913+16. Those estimates suggest that one compact NS/NS binary is born every 10^5 years in our galaxy and that the brightest NS/NS binaries will fall in the indicated region in Fig. 5, extending out to a high-frequency limit of $\simeq 3 \times 10^{-3}$Hz (corresponding to a remaining time to coalescence of 10^5 years). The birth rate might be much higher than $1/10^5$years, according to progenitor evolutionary arguments [46,48,49,50,51,52], in which case LISA would see brighter and higher-frequency binaries than shown in Fig. 5. LISA's observations should easily reveal the true compact NS/NS birth rate and also the birth rates of NS/BH and BH/BH binaries—classes of objects that have not yet been discovered electromagnetically. For further details see [53,23]; for estimates of LISA's angular resolution when observing such binaries, see [54].

6.2 *The Final Inspiral of a NS/NS Binary*

As a result of their loss of orbital energy to gravitational waves, the PSR 1913+16 NS's are gradually spiraling inward at a rate that agrees with general relativity's prediction to within the measurement accuracy (a fraction of a percent) [44]—a remarkable but indirect confirmation that gravitational waves do exist and are correctly described by general relativity. If we wait roughly 10^8 years, this inspiral will bring the waves into the LIGO/ VIRGO high-frequency band. As the NS's continue their inspiral, over a time of about 15 minutes the waves will sweep through the LIGO/VIRGO band, from ~ 10 Hz to $\sim 10^3$ Hz, at which point the NS's will collide and merge. It is this last 15 minutes of inspiral, with $\sim 16,000$ cycles of waveform oscillation, and the final merger, that the LIGO/VIRGO network seeks to monitor.

To what distance must LIGO/VIRGO look, in order to see such inspirals several times per year? Beginning with our galaxy's conservative, pulsar-observation-based NS/NS event rate of one every 100,000 years (Sec. 6.1) and extrapolating out through the Universe, one infers an event rate of several per year at 200 Mpc [45,46,47]. If arguments based on simulations of binary evolution are correct [46,48,49,50,51,52] (Sec. 6.1), the distance for several per year could be as small as 23 Mpc—though such a small distance entails stretching all the numbers to near their breaking point of plausibility [46]. If one stretches all numbers to the opposite, most pessimistic extreme, one infers several per year at 1000 Mpc [46]. Whatever may be the true distance for several per year, once LIGO/VIRGO reaches that distance, each further improvement of sensitivity by a factor 2 will increase the observed event rate by $2^3 \simeq 10$.

Figure 3 compares the projected LIGO sensitivities [12] with the wave strengths from NS/NS inspirals at various distances from Earth. From that comparison we see that LIGO's first interferometers can reach 30Mpc, where the most extremely optimistic estimates predict several per year; the enhanced interferometers can reach 300Mpc where the binary-pulsar-based, conservative estimates predict ~ 10 per year; the advanced interferometers can reach 1000Mpc where even the most extremely pessimistic of estimates predict several per year.

6.3 *Inspiral Waveforms and the Information they Carry*

Neutron stars have such intense self gravity that it is exceedingly difficult to deform them. Correspondingly, as they spiral inward in a compact binary, they do not gravitationally deform each other significantly until several orbits before their final coalescence [55,56]. This means that the inspiral waveforms are determined to high accuracy by only a few, clean parameters: the masses

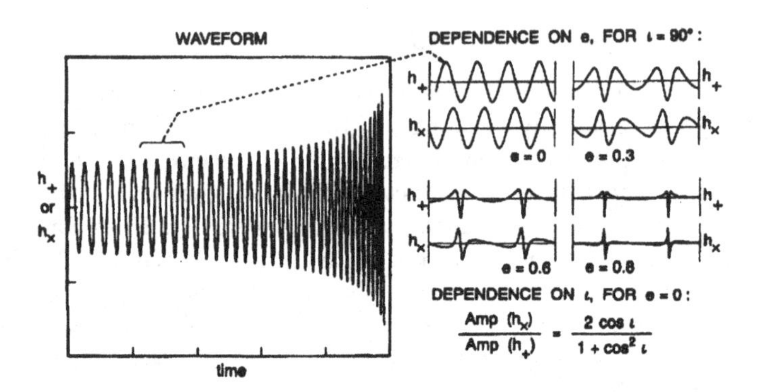

Figure 6. Waveforms from the inspiral of a compact binary (NS/NS, NS/BH, or BH/BH), computed using Newtonian gravity for the orbital evolution and the quadrupole-moment approximation for the wave generation. (From Ref. [12].)

and spin angular momenta of the stars, and the initial orbital elements (i.e. the elements when the waves enter the LIGO/VIRGO band). The same is true for NS/BH and BH/BH binaries. The following description of inspiral waveforms is independent of whether the binary's bodies are NS's or BH's.

Though tidal deformations are negligible during inspiral, relativistic effects can be very important. If, for the moment, we ignore the relativistic effects—i.e., if we approximate gravity as Newtonian and the wave generation as due to the binary's oscillating quadrupole moment [11], then the shapes of the inspiral waveforms $h_+(t)$ and $h_\times(t)$ are as shown in Fig. 6.

The left-hand graph in Fig. 6 shows the waveform increasing in amplitude and sweeping upward in frequency (i.e., undergoing a "chirp") as the binary's bodies spiral closer and closer together. The ratio of the amplitudes of the two polarizations is determined by the inclination ι of the orbit to our line of sight (lower right in Fig. 6). The shapes of the individual waves, i.e. the waves' harmonic content, are determined by the orbital eccentricity (upper right). (Binaries produced by normal stellar evolution should be highly circular due to past radiation reaction forces, but compact binaries that form by capture events, in dense star clusters that might reside in galactic nuclei [57], could be quite eccentric.) If, for simplicity, the orbit is circular, then the rate at which the frequency sweeps or "chirps", df/dt [or equivalently the number of cycles spent near a given frequency, $n = f^2(df/dt)^{-1}$] is determined solely, in the Newtonian/quadrupole approximation, by the binary's so-called *chirp mass*, $M_c \equiv (M_1 M_2)^{3/5}/(M_1+M_2)^{1/5}$ (where M_1 and M_2 are the two bodies' masses). The amplitudes of the two waveforms are determined by the chirp mass, the distance to the source, and the orbital inclination. Thus (in the Newtonian/quadrupole approximation), by measuring the two amplitudes, the frequency sweep, and the harmonic content of the inspiral waves, one can determine as direct, resulting observables, the source's distance, chirp mass, inclination, and eccentricity [58,30]. (For binaries at cosmological distances, the observables are the "luminosity distance," "redshifted" chirp mass $(1+z)M_c$, inclination, and eccentricity; cf. Sec. 7.2.)

As in binary pulsar observations [44], so also here, relativistic effects add further information: they influence the rate of frequency sweep and produce waveform modulations in ways that depend on the binary's dimensionless ratio $\eta = \mu/M$ of reduced mass $\mu = M_1 M_2/(M_1+M_2)$ to total mass $M = M_1+M_2$ and on the spins of the binary's two bodies. These relativistic effects are reviewed and discussed at length in Refs. [59,60]. Two deserve special mention: (i) As the waves emerge from the binary, some of them get backscattered one or more times off the binary's spacetime curvature, producing wave *tails*. These tails act back on the binary, modifying its radiation reaction force and thence its inspiral rate in a measurable way. (ii) If the orbital plane is inclined to one or both of the binary's spins, then the spins drag inertial frames in the binary's vicinity (the "Lense-Thirring effect"), this frame dragging causes the orbit to precess, and the precession modulates the waveforms [59,61,62].

Remarkably, the relativistic corrections to the frequency sweep—tails, spin-induced precession and others—will be measurable with rather high accuracy, even though they are typically $\lesssim 10$ per cent of the Newtonian contribution, and even though the typical signal to noise ratio will be only ~ 9. The reason is as follows [63,64,59]:

The frequency sweep will be monitored by the method of "matched filters"; in other words, the incoming, noisy signal will be cross correlated with theoretical templates. If the signal and the templates gradually get out of phase with each other by more than $\sim 1/10$ cycle as the waves sweep through the LIGO/VIRGO band, their cross correlation will be significantly reduced. Since the total number of cycles spent in the LIGO/VIRGO band will be $\sim 16{,}000$ for a NS/NS binary, ~ 3500 for NS/BH, and ~ 600 for BH/BH, this means that LIGO/VIRGO should be able to measure the frequency sweep to a fractional precision $\lesssim 10^{-4}$, compared to which the relativistic effects are very large. (This is essentially the same method as Joseph Taylor and colleagues use for high-accuracy radio-wave measurements of relativistic effects in binary pulsars [44].)

Analyses using the theory of optimal signal processing predict the following typical accuracies for LIGO/VIRGO measurements based solely on the frequency sweep (i.e., ignoring modulational information) [65]: (i) The chirp mass M_c will typically be measured, from the Newtonian part of the frequency sweep, to $\sim 0.04\%$ for a NS/NS binary and $\sim 0.3\%$ for a system containing at least one BH. (ii) *If* we are confident (e.g., on a statistical basis from measurements of many previous binaries) that the spins are a few percent or less of the maximum physically allowed, then the reduced mass μ will be measured to $\sim 1\%$ for NS/NS and NS/BH binaries, and $\sim 3\%$ for BH/BH binaries. (Here and below NS means a $\sim 1.4M_\odot$ neutron star and BH means a $\sim 10M_\odot$ black hole.) (iii) Because the frequency dependences of the (relativistic) μ effects and spin effects are not sufficiently different to give a clean separation between μ and the spins, if we have no prior knowledge of the spins, then the spin/μ correlation will worsen the typical accuracy of μ by a large factor, to $\sim 30\%$ for NS/NS, $\sim 50\%$ for NS/BH, and a factor ~ 2 for BH/BH. These worsened accuracies might be improved somewhat by waveform modulations caused by the spin-induced precession of the orbit [61,62], and even without modulational information, a certain combination of μ and the spins will be determined to a few per cent. Much additional theoretical work is needed to firm up the measurement accuracies.

To take full advantage of all the information in the inspiral waveforms will require theoretical templates that are accurate, for given masses and spins, to a fraction of a cycle during the entire sweep through the LIGO/VIRGO band. Such templates are being computed by an international consortium of relativity theorists (Blanchet and Damour in France, Iyer in India, Will and Wiseman in the U.S.) [66], using post-Newtonian expansions of the Einstein field equations, of the sort pioneered by Chandrasekhar [67,68]. This enterprise is rather like computing the Lamb shift to high order in powers of the fine structure constant, for comparison with experiment and testing of quantum electrodynamics. Cutler and Flanagan [69] have estimated the order to which the computations must be carried in order that systematic errors in the theoretical templates will not significantly impact the information extracted from the LIGO/VIRGO observational data. The answer appears daunting: radiation-reaction effects must be computed to three full post-Newtonian orders [six orders in v/c =(orbital velocity)/(speed of light)] beyond Chandra's leading-order radiation reaction, which itself is 5 orders in v/c beyond the Newtonian theory of gravity, so the required calculations are $O[(v/c)^{6+5}] = O[(v/c)^{11}]$. By clever use of Padé approximates, these requirements might be relaxed [70].

In the late 1960's, when Chandra and I were first embarking on our respective studies of gravitational waves, Chandra set out to compute the first 5 orders in v/c beyond Newton, i.e., in his own words, "to solve Einstein's equations through the 5/2 post-Newtonian", thereby fully understanding leading-order radiation reaction and all effects leading up to it. Some colleagues thought his project not worth the enormous personal effort that he put into it. But Chandra was prescient. He had faith in the importance of his effort, and history has proved him right. The results of his "5/2 post-Newtonian" [68] calculation have now been verified to accuracy better than 1% by observations of the inspiral of PSR 1913+16; and the needs of LIGO/VIRGO data analysis are now driving the calculations onward from $O[(v/c)^5]$ to $O[(v/c)^{11}]$. This epitomizes a major change in the field of relativity research: At last, 80 years after Einstein formulated general relativity, experiment has become a major driver for theoretical analyses.

Remarkably, the goal of $O[(v/c)^{11}]$ is achievable. The most difficult part of the computation, the radiation reaction, has been evaluated to $O[(v/c)^9]$ beyond Newton by the French/Indian/American consortium [66] and $O[(v/c)^{11}]$ is now being pursued.

These high-accuracy waveforms are needed only for extracting information from the inspiral waves after the waves have been discovered; they are not needed for the discovery itself. The discovery is best achieved using a different family of theoretical waveform templates, one that covers the space of potential waveforms in a manner that minimizes computation time instead of a manner that ties quantitatively into general relativity theory [59,71]. Such templates are under development.

6.4 NS/NS Merger Waveforms and their Information

The final merger of a NS/NS binary should produce waves that are sensitive to the equation of state of nuclear matter, so such mergers have the potential to teach us about the nuclear equation of state [12,59]. In essence, LIGO/VIRGO will be studying nuclear physics via the collisions of atomic nuclei that have nucleon numbers $A \sim 10^{57}$—somewhat larger than physicists are normally accustomed to. The accelerator used to drive these "nuclei" up to half the speed of light is the binary's self gravity, and the radiation by which the details of the collisions are probed is gravitational.

Unfortunately, the NS/NS merger will emit its gravitational waves in the kHz frequency band ($600\text{Hz} \lesssim f \lesssim 2500\text{Hz}$) where photon shot noise will prevent the waves from being studied by the standard, "workhorse," broad-band interferometers of Fig. 3. However, it may be possible to measure the waves and extract their equation-of-state information using a "xylophone" of specially configured narrow-band detectors (signal-recycled or resonant-sideband-extraction interferometers, and/or spherical or icosahedral resonant-mass detectors; Sec. 3.3 and Refs. [59,72]). Such measurements will be very difficult and are likely only when the LIGO/VIRGO network has reached a mature stage.

A number of research groups [73] are engaged in numerical simulations of NS/NS mergers, with the goal not only to predict the emitted gravitational waveforms and their dependence on equation of state, but also (more immediately) to learn whether such mergers might power the γ-ray bursts that have been a major astronomical puzzle since their discovery in the early 1970s.

NS/NS mergers are a promising explanation for γ-ray bursts because (i) some bursts are known, from intergalactic absorption lines, to come from cosmological distances [74], (ii) the bursts have a distribution of number versus intensity that suggests most lie at near-cosmological distances, (iii) their event rate is roughly the same as that conservatively estimated for NS/NS mergers ($\sim$ 1000 per year out to cosmological distances; a few per year at 300Mpc); and (iv) it is plausible that the final NS/NS merger will create a γ-emitting fireball with enough energy to account for the bursts [75,76]. If enhanced LIGO interferometers were now in operation and observing NS/NS inspirals, they could report definitively whether or not the γ-bursts are produced by NS/NS binaries; and if the answer were yes, then the combination of γ-burst data and gravitational-wave data could bring valuable information that neither could bring by itself. For example, it would reveal when, to within a few msec, the γ-burst is emitted relative to the moment the NS's first begin to touch; and by comparing the γ and gravitational times of arrival, we could test whether gravitational waves propagate with the speed of light to a fractional precision of $\sim 0.01\text{sec}/10^9\,\text{lyr} = 3 \times 10^{-19}$.

6.5 NS/BH Mergers

A neutron star (NS) spiraling into a black hole of mass $M \gtrsim 10M_\odot$ should be swallowed more or less whole. However, if the BH is less massive than roughly $10M_\odot$, and especially if it is rapidly rotating, then the NS will tidally disrupt before being swallowed. Little is known about the disruption and accompanying waveforms. To model them with any reliability will likely require full numerical relativity, since the circumferences of the BH and NS will be comparable and their physical separation at the moment of disruption will be of order their separation. As with NS/NS, the merger waves should carry equation of state information and will come out in the kHz band, where their detection will require advanced, specialty detectors.

7 Black Hole Binaries

7.1 BH/BH Inspiral, Merger, and Ringdown

We turn, next, to binaries made of two black holes with comparable masses (BH/BH binaries). The LIGO/VIRGO network can detect and study waves from the last few minutes of the life of such a binary if its total mass is $M \lesssim 1000M_\odot$ ("stellar-mass black holes"), cf. Fig. 3; and LISA can do the same for the mass range $1000M_\odot \lesssim M \lesssim 10^8 M_\odot$ ("supermassive black holes"), cf. Fig. 5.

The timescales for the binary's dynamics and its waveforms are proportional to its total mass M. All other aspects of the dynamics and waveforms, after time scaling, depend solely on quantities that are dimensionless in geometrized units ($G = c = 1$): the ratio of the two BH masses, the BH spins divided by the squares of their masses, etc. Consequently, the black-hole physics to be studied is the same for supermassive holes in LISA's low-frequency band as for stellar-mass holes in LIGO/VIRGO's high-frequency band. LIGO/VIRGO is likely to make moderate-accuracy studies of this physics; and LISA, flying later, can achieve high accuracy.

The binary's dynamics and its emitted waveforms can be divided into three epochs: ***inspiral, merger, and ringdown*** [77]. The inspiral epoch terminates when the holes reach their last stable orbit and begin plunging toward each other. The merger epoch lasts from the beginning of plunge until the holes have merged and can be regarded as a single hole undergoing large-amplitude, quasinormal-mode vibrations. In the ringdown epoch, the hole's vibrations decay due to wave emission, leaving finally a quiescent, spinning black hole.

The inspiral epoch has been well studied theoretically using post-Newtonian expansions (Sec. 6.3), except for the last factor ~ 3 of upward frequency sweep, during which the post-Newtonian expansions may fail. The challenge of computing this last piece of the inspiral is called the "intermediate binary black hole problem" (IBBH) and is a subject of current research in my own group and elsewhere. The merger epoch can be studied theoretically only via supercomputer simulations. Techniques for such simulations are being developed by several research groups, including an eight-university American consortium of numerical relativists and computer scientists called the Binary Black Hole Grand Challenge Alliance [78]. Chandrasekhar and Detweiler [79,80] pioneered the study of the ringdown epoch using the Teukolsky formalism for first-order perturbations of spinning (Kerr) black holes (see Chandra's classic book [81]), and the ringdown is now rather well understood except for the strengths of excitation of the various vibrational modes, which the merger observations and computations should reveal.

The merger epoch, as yet, is very poorly understood. We can expect it to consist of large-amplitude, highly nonlinear vibrations of spacetime curvature—a phenomenon of which we have very little theoretical understanding today. Especially fascinating will be the case of two spinning black holes whose spins are not aligned with each other or with the orbital angular momentum. Each of the three angular momentum vectors (two spins, one orbital) will drag space in its vicinity into a tornado-like swirling motion—the general relativistic "dragging of inertial frames"—so the binary is rather like two tornados with orientations skewed to each other, embedded inside a third, larger tornado with a third orientation. The dynamical evolution of such a complex configuration of coalescing spacetime warpage, as revealed by its emitted waves, might bring us surprising new insights into relativistic gravity [12].

7.2 BH/BH Signal Strengths and Detectability

Flanagan and Hughes [77] have recently estimated the signal strengths produced in LIGO and in LISA by the waves from equal-mass BH/BH binaries for each of the three epochs, inspiral, merger, and ringdown; and along with signal strengths, they have estimated the distances to which LIGO and LISA can detect the waves. In their estimates, Flanagan and Hughes make plausible assumptions about the waves' unknown aspects. The estimated signal strengths are shown in Fig. 7 for the first LIGO interferometers, Fig. 8 for advanced LIGO interferometers, and Fig. 9 for LISA. Because LIGO and LISA can both reach out to cosmological distances, these figures are drawn in a manner that includes cosmological effects: they are valid for any homogeneous, isotropic model of our universe. This is achieved by plotting observables that are extracted from the measured waveforms: the binary's "redshifted" total mass $(1+z)M$ on the horizontal axis (where z is the source's cosmological redshift) and its "luminosity distance" [82] on the right axis. The signal-to-noise ratio (left axis) scales inversely with the luminosity distance.

We have no good observational handle on the coalescence rate of stellar-mass BH/BH binaries. However, for BH/BH binaries with total mass $M \sim 5$ to $50 M_\odot$ that arise from ordinary main-sequence progenitors, estimates based on the progenitors' birth rates and on simulations of their subsequent evolution suggest a coalescence rate in our galaxy of one per (1 to 30) million years [51,48]. These rough estimates imply that to see one coalescence per year with $M \sim 5$ to $50 M_\odot$, LIGO/VIRGO must reach out to a distance $\sim$ (300 to 900) Mpc. Other plausible scenarios (e.g. BH/BH binary formation in

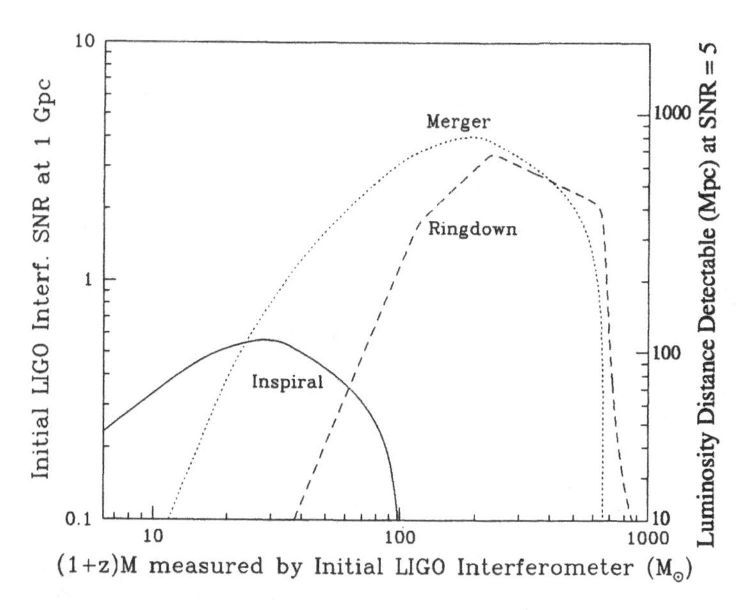

Figure 7. The inspiral, merger, and ringdown waves from equal-mass black-hole binaries as observed by LIGO's initial interferometers: The luminosity distance to which the waves are detectable (right axis) and the signal-to-noise ratio for a binary at 1Gpc (left axis), as functions of the binary's redshifted total mass (bottom axis). (Figure adapted from Flanagan and Hughes [77].)

dense stellar clusters that reside in globular clusters and galactic nuclei [57]) could produce higher event rates and larger masses, but little reliable is known about them (cf. Sec. I.A.ii of [77]).

For comparison, the first LIGO interferometers can reach 300Mpc for $M = 50M_\odot$ but only 40Mpc for $M = 5M_\odot$ (Fig. 7); enhanced interferometers can reach about 10 times farther, and advanced interferometers about 30 times farther (Fig. 8. These numbers suggest that (i) if waves from BH/BH coalescences are not detected by the first LIGO/VIRGO interferometers, they are likely to be detected along the way from the first interferometers to the enhanced; and (ii) BH/BH coalescences might be detected sooner than NS/NS coalescences (cf. Sec. 6.2).

For binaries with $M(1+z) \gtrsim 40M_\odot$, the highly interesting merger signal should be stronger than the inspiral signal, and for $M \gtrsim 100M_\odot$, the ringdown should be stronger than inspiral (Fig. 7). Thus, it may well be that early in the life of the LIGO/VIRGO network, observers and theorists will be struggling

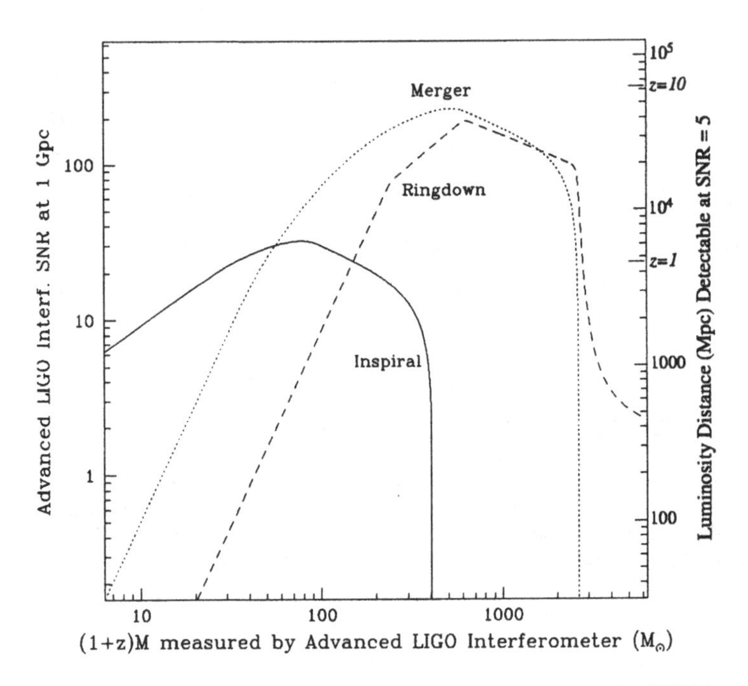

Figure 8. The waves from equal-mass black-hole binaries as observed by LIGO's advanced interferometers; cf. the caption for Fig. 7. On the right side is shown not only the luminosity distance to which the signals can be seen (valid for any homogeneous, isotropic cosmology), but also the corresponding cosmological redshift z, assuming vanishing cosmological constant, a spatially flat universe, and a Hubble constant $H_o = 75$ km/s/Mpc. (Figure adapted from Flanagan and Hughes [77].)

to understand the merger of binary black holes by comparison of computed and observed waveforms.

LIGO's advanced interferometers (Fig. 8) can see the merger waves, for

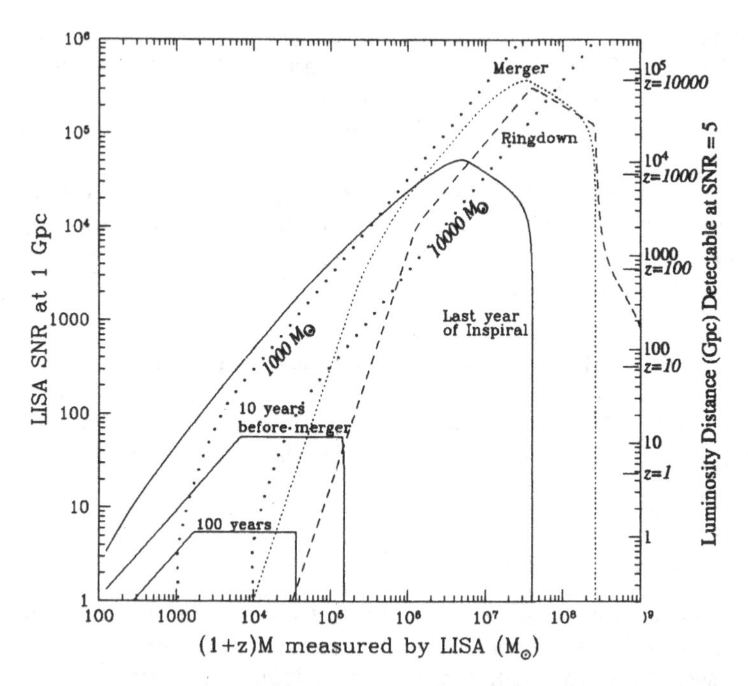

Figure 9. The waves from equal-mass, supermassive black-hole binaries as observed by LISA in one year of integration time; cf. the captions for Figs. 7 and 8. The wide-spaced dots are curves of constant binary mass M, for use with the right axis, assuming vanishing cosmological constant, a spatially flat universe, and a Hubble constant $H_o = 75$ km/s/Mpc. The bottom-most curves are the signal strengths after one year of signal integration, for BH/BH binaries 10 years and 100 years before their merger. (Figure adapted from Flanagan and Hughes [77].)

$20M_\odot \lesssim M \lesssim 200M_\odot$) out to a cosmological redshift $z \simeq 5$; and for binaries at $z = 1$ in this mass range, they can achieve a signal to noise ratio (assuming optimal signal processing [77]) of about 25 in each interferometer.

While these numbers are impressive, they pale by comparison with LISA (Fig. 9), which can detect the merger waves for $1000M_\odot \lesssim M \lesssim 10^5 M_\odot$ out to redshifts $z \sim 3000$ (far earlier in the life of the universe than the era when the first supermassive black holes are likely to have formed). Correspondingly, LISA can achieve signal to noise ratios of thousands for mergers with $10^5 \lesssim M \lesssim 10^8 M_\odot$ at redshifts of order unity, and from the inspiral waves can infer the binary's parameters (redshifted masses, luminosity distance, direction, ...) with high accuracy [54].

Unfortunately, it is far from obvious whether the event rate for such supermassive BH/BH coalescences will be interestingly high. Conservative estimates suggest a rate of ~ 0.1/yr, while plausible scenarios for aspects of the universe about which we are rather ignorant can give rates as high as 1000/yr [83].

If the coalescence rate is only 0.1/yr, then LISA should still see ~ 3 BH/BH binaries with $3000 M_\odot \lesssim M \lesssim 10^5 M_\odot$ that are ~ 30 years away from their final merger. These slowly inspiraling binaries should be visible, with one year of integration, out to a redshift $z \sim 1$ (bottom part of Fig. 9).

8 Payoffs from Binary Coalescence Observations

Among the scientific payoffs that should come from LIGO/VIRGO's and/or LISA's observations of binary coalescence are the following; others have been discussed above.

8.1 Christodoulou Memory

As the gravitational waves from a binary's coalescence depart from their source, the waves' energy creates (via the nonlinearity of Einstein's field equations) a secondary wave called the "Christodoulou memory" [84,85,86]. This memory, arriving at Earth, can be regarded rigorously as the combined gravitational field of all the gravitons that have been emitted in directions other than toward the Earth [85]. The memory builds up on the timescale of the primary energy emission profile, and grows most rapidly when the primary waves are being emitted most strongly: during the end of inspiral and the merger. Unfortunately, the memory is so weak that in LIGO only advanced interferometers have much chance of detecting and studying it—and then, only for BH/BH coalescences and not for NS/NS [87]. LISA, by contrast, should easily be able to measure the memory from supermassive BH/BH coalescences.

8.2 Testing General Relativity

Corresponding to the very high post-Newtonian order to which a binary's inspiral waveforms must be computed for use in LIGO/VIRGO and LISA data analysis (Sec. 6.3), measurements of the inspiral waveforms can be used to test general relativity with very high accuracy. For example, in scalar-tensor theories (some of which are attractive alternatives to general relativity [88]), radiation reaction due to emission of scalar waves places a unique signature on the measured inspiral waveforms—a signature that can be searched for with high precision [89]. Similarly, the inspiral waveforms can be used to measure with high accuracy several fascinating general relativistic phenomena in addition to the Christodoulou memory: the influence of the tails of the emitted waves on radiation reaction in the binary (Sec. 6.3), the Lens-Thirring orbital precision induced by the binary's spins (Sec. 6.3), and a unique relationship among the multipole moments of a quiescent black hole which is dictated by a hole's "two-hair theorem" (Sec. 8.4).

The ultimate test of general relativity will be detailed comparisons of the predicted and observed waveforms from the highly nonlinear spacetime-warpage vibrations of BH/BH mergers (Sec. 7.1).

8.3 Cosmological Measurements

Binary inspiral waves can be used to measure the Universe's Hubble constant, deceleration parameter, and cosmological constant [58,30,90,91]. The keys to such measurements are that: (i) Advanced interferometers in LIGO/VIRGO will be able to see NS/NS inspirals out to cosmological redshifts $z \sim 0.3$, and NS/BH out to $z \sim 2$. (ii) The direct observables that can be extracted from the observed waveforms include a source's luminosity distance (measured to an accuracy ~ 10 per cent in a large fraction of cases), and its direction on the sky (to accuracy ~ 1 square degree)—accuracies good enough that only one or a few electromagnetically-observed clusters of galaxies should fall within the 3-dimensional gravitational error boxes. This should make possible joint gravitational/electromagnetic statistical studies of our Universe's magnitude-redshift relation, with gravity giving luminosity distances and electromagnetism giving the redshifts [58,30]. (iii) Another direct gravitational observable is any redshifted mass $(1+z)M$ in the system. Since the masses of NS's in binaries seem to cluster around $1.4M_\odot$, measurements of $(1+z)M$ can provide a handle on the redshift, even in the absence of electromagnetic aid; so gravitational-wave observations alone may be used, in a statistical way, to measure the magnitude-redshift relation [90,91].

LISA, with its ability to detect BH/BH binaries with $M \sim 1000$ to $100,000 M_\odot$ out to redshifts of thousands, could search for the earliest epochs of supermassive black hole activity—if the Universe is kind enough to grant us a large event rate.

8.4 Mapping Quiescent Black Holes; Searching for Exotic Relativistic Bodies

Ryan [92] has shown that, when a white dwarf, neutron star or small black hole spirals into a much more massive, compact central body, the inspiral waves will carry a "map" of the massive body's external spacetime geometry. Since the body's spacetime geometry is uniquely characterized by the values of the body's multiple moments, we can say equivalently that the inspiral waves carry, encoded in themselves, the values of all the body's multipole moments.

By measuring the inspiral waveforms and extracting their map (i.e., measuring the lowest few multipole moments), we can determine whether the massive central body is a black hole or some other kind of exotic compact object [92]; see below.

The inspiraling object's orbital energy E at fixed frequency f (and correspondingly at fixed orbital radius a) scales as $E \propto \mu$, where μ is the object's mass; the gravitational-wave luminosity $\dot{E}$ scales as $\dot{E} \propto \mu^2$; and the time to final merger thus scales as $t \sim E/\dot{E} \propto 1/\mu$. This means that the smaller is μ/M (where M is the central body's mass), the more orbits are spent in the central body's strong-gravity region, $a \lesssim 10GM/c^2$, and thus the more detailed and accurate will be the map of the body's spacetime geometry encoded in the emitted waves.

For holes observed by LIGO/VIRGO, the most extreme mass ratio that we can hope for is $\mu/M \sim 1M_\odot/300M_\odot$, since for $M > 300M_\odot$ the inspiral waves are pushed to frequencies below the LIGO/VIRGO band. This limit on μ/M seriously constrains the accuracy with which LIGO/VIRGO can hope to map the spacetime geometry. A detailed study by Ryan [93] (but one that is rather approximate because we do not know the full details of the waveforms) suggests that LIGO/VIRGO might *not* be able to distinguish cleanly between quiescent black holes and other types of massive central bodies.

By contrast, LISA can observe the final inspiral waves from objects of any mass $\mu \gtrsim 1M_\odot$ spiraling into central bodies of mass $3 \times 10^5 M_\odot \lesssim M \lesssim 3 \times 10^7 M_\odot$ out to 3Gpc. Figure 5 shows the example of a $10M_\odot$ black hole spiraling into a $10^6 M_\odot$ black hole at 3Gpc distance. The inspiral orbit and waves are strongly influenced by the hole's spin. Two cases are shown [94]: an inspiraling circular orbit around a non-spinning hole, and a prograde, circular, equatorial orbit around a maximally spinning hole. In each case the dot at the upper left end of the arrowed curve is the frequency and characteristic amplitude one year before the final coalescence. In the nonspinning case, the small hole spends its last year spiraling inward from $r \simeq 7.4GM/c^2$ (3.7 Schwarzschild radii) to its last stable circular orbit at $r = 6GM/c^2$ (3 Schwarzschild radii). In the maximal spin case, the last year is spent traveling from $r = 6GM/c^2$ (3 Schwarzschild radii) to the last stable orbit at $r = GM/c^2$ (half a Schwarzschild radius). The $\sim 10^5$ cycles of waves during this last year should carry, encoded in themselves, rather accurate values for the massive hole's lowest few multipole moments [92,93] (or, equivalently, a rather accurate map of the hole's spacetime geometry).

If the measured moments satisfy the black-hole "two-hair" theorem (usually incorrectly called the "no-hair" theorem), i.e. if they are all determined uniquely by the measured mass and spin in the manner of the Kerr metric, then we can be sure the central body is a black hole. If they violate the two-hair theorem, then (assuming general relativity is correct), either the central body was an exotic object—e.g. a spinning boson star which should have three "hairs" [95], a soliton star [96] or a naked singularity—rather than a black hole, or else an accretion disk or other material was perturbing its orbit [97]. From the evolution of the waves one can hope to determine which is the case, and to explore the properties of the central body and its environment [98].

Models of galactic nuclei, where massive holes (or other massive central bodies) reside, suggest that inspiraling stars and small holes typically will be in rather eccentric orbits [99,100]. This is because they get injected into such orbits via gravitational deflections off other stars, and by the time gravitational radiation reaction becomes the dominant orbital driving force, there is not enough inspiral left to strongly circularize their orbits. Such orbital eccentricity will complicate the waveforms and complicate the extraction of information from them. Efforts to understand the emitted waveforms, for central bodies with arbitrary multipole moments, are just now getting underway [92,101]. Even for central black holes, those efforts are at an early stage; for example, only recently have we learned how to compute the influence of radiation reaction on inspiraling objects in fully relativistic, nonequatorial orbits around a black hole [102,103].

The event rates for inspiral into supermassive black holes (or other supermassive central bodies) are not well understood. However, since a significant fraction of all galactic nuclei are thought to contain supermassive holes, and since white dwarfs and neutron stars, as well as small black holes, can withstand tidal disruption as they plunge toward a supermassive hole's horizon, and since LISA can see inspiraling bodies as small as $\sim 1M_\odot$ out to 3Gpc distance, the event rate is likely to be interestingly large. Sigurdsson and Rees give a "very conservative" estimate of one inspiral event per year within 1Gpc distance, and 100–1000 sources detectable by LISA at lower frequencies "en route" toward their final plunge.

9 Conclusion

It is now 37 years since Joseph Weber initiated his pioneering development of gravitational-wave detectors [1], 26 years since Robert Forward [104] and Rainer Weiss [2] initiated work on interferometric detectors, and about 35 years since

Chandra and others launched the modern era of theoretical research on relativistic stars and black holes. Since then, hundreds of talented experimental physicists have struggled to improve the sensitivities of gravitational-wave detectors, and hundreds of theorists have explored general relativity's predictions for stars and black holes.

These two parallel efforts are now intimately intertwined and are pushing toward an era in the not distant future, when measured gravitational waveforms will be compared with theoretical predictions to learn how many and what kinds of relativistic objects *really* populate our Universe, and how these relativistic objects *really* are structured and *really* behave when quiescent, when vibrating, and when colliding.

Acknowledgments

My group's research on gravitational waves and their relevance to LIGO/VIRGO and LISA is supported in part by NSF grants AST-9731698 and PHY-9424337 and by NASA grant NAG5-6840. This article is a slightly revised and updated version of my Ref. [105], and large portions of it have been adapted and updated from my Ref. [106].

References

1. J. Weber. *Phys. Rev.*, 117:306, 1960.
2. R. Weiss. *Quarterly Progress Report of RLE, MIT*, 105:54, 1972.
3. M. S. Turner. *Phys. Rev. D*, 55:R435, 1997.
4. L. P. Grishchuk. *Phys. Rev. D*, 53:6784, 1996.
5. M. Gasperini, M. Giovannini, and G. Veneziano. *Phys. Rev. D*, 52:R6651, 1995.
6. V. M. Kaspi, J. H. Taylor, and M. F. Ryba. *Astrophys. J.*, 428:713, 1994.
7. Ya. B. Zel'dovich. *Mon. Not. Roy. Astron. Soc.*, 192:663, 1980.
8. A. Vilenkin. *Phys. Rev. D*, 24:2082, 1981.
9. A. Kosowsky and M. S. Turner. *Phys. Rev. D*, 49:2837, 1994.
10. X. Martin and A. Vilenkin. *Phys. Rev. Lett.*, 77:2879, 1996.
11. K. S. Thorne. In S. W. Hawking and W. Israel, editors, *Three Hundred Years of Gravitation*, pages 330–458. Cambridge University Press, 1987.
12. A. Abramovici et. al. *Science*, 256:325, 1992.
13. A. Abramovici et. al. *Physics Letters A*, 218:157, 1996.
14. C. Bradaschia et. al. *Nucl. Instrum. & Methods*, A289:518, 1990.
15. B. Barish and G. Sanders et. al. LIGO Advanced R and D Program Proposal, Caltech/MIT, unpublished, 1996.

16. J. Hough and K. Danzmann et. al. GEO600, Proposal for a 600 m Laser-Interferometric Gravitational Wave Antenna, unpublished, 1994.
17. K. Kuroda et. al. In I. Ciufolini and F. Fidecaro, editors, *Gravitational Waves: Sources and Detectors*, page 100. World Scientific, 1997.
18. B. J. Meers. *Phys. Rev. D*, 38:2317, 1988.
19. M. J. Mizuno, K. A. Strain, P. G. Nelson, J. M. Chen, R. Schilling, A. Rudiger, W. Winkler, and K. Danzman. *Phys. Lett. A*, 175:273, 1993.
20. W. W. Johnson and S. M. Merkowitz. *Phys. Rev. Lett.*, 70:2367, 1993.
21. G. Frossati. *J. Low Temp. Phys.*, 101:81, 1995.
22. G. V. Pallottino. In I. Ciufolini and F. Fidecaro, editors, *Gravitational Waves: Sources and Detectors*, page 159. World Scientific, 1997.
23. P. Bender et. al. *LISA, Laser interferometer space antenna for the detection and observation of gravitational waves: Pre-Phase A Report.* Max-Planck-Institut für Quantenoptik, MPQ 208, December 1995.
24. A. G. Petschek, editor. *Supernovae.* Springer Verlag, 1990.
25. H. A. Bethe. *Rev. Mod. Phys*, 62:801, 1990.
26. A. Burrows, 1994. Private communication.
27. E. Müller and H.-T. Janka. *Astron. Astrophys.*, 317:140, 1997.
28. L. S. Finn. *Ann. N. Y. Acad. Sci.*, 631:156, 1991.
29. R. Mönchmeyer, G. Schäfer, E. Müller, and R. E. Kates. *Astron. Astrophys.*, 256:417, 1991.
30. B. F. Schutz. *Class. Quant. Grav.*, 6:1761, 1989.
31. D. Lai and S. L. Shapiro. *Astrophys. J.*, 442:259, 1995.
32. J. L. Houser, J. M. Centrella, and S. C. Smith. *Phys. Rev. Lett.*, 72:1314, 1994.
33. I. A. Bonnell and J. E. Pringle. *Mon. Not. Roy. Astron. Soc.*, 273:L12, 1995.
34. M. Zimmermann and E. Szedenits. *Phys. Rev. D*, 20:351, 1979.
35. S. L. Shapiro and S. A. Teukolsky. Wiley: Interscience, 1983. Section 10.10 and references cited therein.
36. P. Brady, T. Creighton, C. Cutler, and B. Schutz. *Phys. Rev. D*, 57:2101, 1998.
37. B. F. Schutz, 1995. Private communication.
38. S. Chandrasekhar. *Phys. Rev. Lett.*, 24:611, 1970.
39. J. L. Friedman and B. F. Schutz. *Astrophys. J.*, 222:281, 1978.
40. R. V. Wagoner. *Astrophys. J.*, 278:345, 1984.
41. L. Lindblom. *Astrophys. J.*, 438:265, 1995.
42. L. Lindblom and G. Mendell. *Astrophys. J.*, 444:804, 1995.
43. R. A. Hulse and J. H. Taylor. *Astrophys. J.*, 324:355, 1975.

44. J. H. Taylor. *Rev. Mod. Phys.*, 66:711, 1994.
45. R. Narayan, T. Piran, and A. Shemi. *Astrophys. J.*, 379:L17, 1991.
46. E. S. Phinney. *Astrophys. J.*, 380:L17, 1991.
47. E. P. J. van den Heuvel and D. R. Lorimer. *Mon. Not. Roy. Astron. Soc.*, 283:L37, 1996.
48. A. V. Tutukov and L. R. Yungelson. *Mon. Not. Roy. Astron. Soc.*, 260:675, 1993.
49. H. Yamaoka, T. Shigeyama, and K. Nomoto. *Astron. Astrophys.*, 267:433, 1993.
50. V. M. Lipunov, K. A. Postnov, and M. E. Prokhorov. *Astrophys. J.*, 423:L121, 1994. And related, unpublished work.
51. V. M. Lipunov, K. A. Postnov, and M. E. Prokhorov. *New Astronomy*, 2:43, 1997.
52. S. F. P. Zwart and H. N. Spreeuw. *Astron. Astrophys.*, 312:670, 1996.
53. D. Hils, P. Bender, and R. F. Webbink. *Astrophys. J.*, 360:75, 1990.
54. C. Cutler. *Phys. Rev. D*, 1998. Submitted; gr/qc-9703068.
55. C. Kochanek. *Astrophys. J.*, 398:234, 1992.
56. L. Bildsten and C. Cutler. *Astrophys. J.*, 400:175, 1992.
57. G. Quinlan and S. L. Shapiro. *Astrophys. J.*, 321:199, 1987.
58. B. F. Schutz. *Nature*, 323:310, 1986.
59. C. Cutler, T. A. Apostolatos, L. Bildsten, L. S. Finn, E. E. Flanagan, D. Kennefick, D. M. Markovic, A. Ori, E. Poisson, G. J. Sussman, and K. S. Thorne. *Phys. Rev. Lett.*, 70:1984, 1993.
60. C. M. Will. In M. Sasaki, editor, *Relativistic Cosmology*, page 83. Universal Academy Press, 1994.
61. T. A. Apostolatos, C. Cutler, G. J. Sussman, and K. S. Thorne. *Phys. Rev. D*, 49:6274, 1994.
62. L. E. Kidder. *Phys. Rev. D*, 52:821, 1995.
63. C. Cutler and E. E. Flanagan. *Phys. Rev. D*, 49:2658, 1994.
64. L. S. Finn and D. F. Chernoff. *Phys. Rev. D*, 47:2198, 1993.
65. E. Poisson and C. M. Will. *Phys. Rev. D*, 52:848, 1995.
66. L. Blanchet, T. Damour, B. R. Iyer, C. M. Will, and A. G. Wiseman. *Phys. Rev. Lett.*, 74:3515, 1995.
67. S. Chandrasekhar. *Selected Papers of S. Chandrasekhar, Vol. 5, Relativistic Astrophysics*. U. Chicago Press, 1990.
68. S. Chandrasekhar and F. P. Esposito. *Astrophys. J.*, 160:153, 1970.
69. C. Cutler and E. E. Flanagan. Phys. Rev. D. Paper in preparation.
70. T. Damour and B.S. Sathyaprakash. Research in progress, 1997.
71. B. J. Owen. *Phys. Rev. D*, 53:6749, 1996.
72. S. A. Hughes, D. Kennefick, D. Laurence, and K. S. Thorne. Phys. Rev.

D. In preparation.
73. Z. G. Xing, J. M. Centrella, and S. W. McMillan. *Phys. Rev. D*, 54:7261, 1996. Also references therein.
74. M. R. Metzger et. al. *Nature*, 387:878, 1997.
75. P. Meszaros. *Ann. N.Y. Acad. Sci*, 759:440, 1995.
76. S. E. Woosley. *Ann. N.Y. Acad. Sci*, 759:446, 1995.
77. E. E. Flanagan and S. A. Hughes. *Phys. Rev. D*, 57:4535; *ibid.*, 4566, 1998.
78. Numerical Relativity Grand Challenge Alliance, 1995. References and information on the World Wide Web, http://jean-luc.ncsa.uiuc.edu/GC.
79. S. Chandrasekhar and S. L. Detweiler. *Proc. Roy. Soc. A*, 344:441, 1975.
80. S. L. Detweiler. *Astrophys. J.*, 239:292, 1980.
81. S. Chandrasekhar. *The Mathematical Theory of Black Holes.* Oxford University Press, 1983.
82. E. W. Kolb and M. S. Turner. *The Early Universe.* Addison-Wesley, 1990. Pages 40–43.
83. M. G. Haehnelt. *Mon. Not. Roy. Astron. Soc.*, 269:199, 1994.
84. D. Christodoulou. *Phys. Rev. Lett.*, 67:1486, 1991.
85. K. S. Thorne. *Phys. Rev. D*, 45:520–524, 1992.
86. A. G. Wiseman and C. M. Will. *Phys. Rev. D*, 44:R2945, 1991.
87. D. Kennefick. *Phys. Rev. D*, 50:3587, 1994.
88. T. Damour and K. Nordtvedt. *Phys. Rev. D*, 48:3436, 1993.
89. C. M. Will. *Phys. Rev. D*, 50:6058, 1994.
90. D. Markovic. *Phys. Rev. D*, 48:4738, 1993.
91. D. F. Chernoff and L. S. Finn. *Astrophys. J. Lett.*, 411:L5, 1993.
92. F. D. Ryan. *Phys. Rev. D*, 52:5707, 1995.
93. F. D. Ryan. Accuracy of estimating the multipole moments of a massive body from the gravitational waves of a binary inspiral. *Phys. Rev. D*, 56:1845, 1997.
94. L. S. Finn and K. S. Thorne. *Phys. Rev. D.* In preparation.
95. F. D. Ryan. Spinning boson stars with large self-interaction. *Phys. Rev. D*, 55:6081, 1997.
96. T. D. Lee and Y. Pang. *Physics Reports*, 221:251, 1992.
97. S. K. Chakrabarti. *Phys. Rev. D*, 53:2901, 1996.
98. F. D. Ryan, L. S. Finn, and K. S. Thorne. *Phys. Rev. Lett.* In preparation.
99. D. Hils and P. Bender. *Astrophys. J. Lett*, 445:L7, 1995.
100. S. Sigurdsson and M. J. Rees. *Mon. Not. Roy. Astron. Soc.*, 284:318, 1997.
101. F. D. Ryan. Scalar waves produced by a scalar charge orbiting a massive

body with arbitrary multipole moments. *Phys. Rev. D*, 56:7732, 1997.
102. Y. Mino, M. Sasaki, and T. Tanaka. *Phys. Rev. D*, 55:3497, 1997.
103. T. C. Quinn and R. M. Wald. *Phys. Rev. D*, 56:3381, 1997.
104. G. E. Moss, L. R. Miller, and R. L. Forward. *Applied Optics*, 10:2495, 1971.
105. K. S. Thorne. In R. M. Wald, editor, *Black Holes and Relativistic Stars*, page 41. University of Chicago Press, 1998.
106. K. S. Thorne. In E. W. Kolb and R. Peccei, editors, *Proceedings of the Snowmass 95 Summer Study on Particle and Nuclear Astrophysics and Cosmology*, page 398. World Scientific, 1995.

CHERN-SIMONS SUPERGRAVITIES WITH OFF-SHELL LOCAL SUPERALGEBRAS

RICARDO TRONCOSO AND JORGE ZANELLI*

Centro de Estudios Científicos de Santiago, Casilla 16443, Santiago 9, Chile
and
Departamento de Física, Universidad de Santiago de Chile, Casilla 307, Santiago 2, Chile

A new family of supergravity theories in odd dimensions is presented. The Lagrangian densities are Chern-Simons forms for the connection of a supersymmetric extension of the anti-de Sitter algebra. The superalgebras are the supersymmetric extensions of the AdS algebra for each dimension, thus completing the analysis of van Holten and Van Proeyen, which was valid for $N = 1$ and for $D = 2, 3, 4, mod\ 8$. The Chern-Simons form of the Lagrangian ensures invariance under the gauge supergroup by construction and, in particular, under local supersymmetry. Thus, unlike standard supergravity, the local supersymmetry algebra closes off-shell and without requiring auxiliary fields. The Lagrangian is explicitly given for $D = 5$, 7 and 11. In all cases the dynamical field content includes the vielbein (e^a_μ), the spin connection (ω^{ab}_μ), N gravitini (ψ^i_μ), and some extra bosonic "matter" fields which vary from one dimension to another. The superalge bras fall into three families: $osp(m|N)$ for $D = 2, 3, 4$, mod 8, $osp(N|m)$ for $D = 6, 7, 8$, mod 8, and $su(m-2, 2|N)$ for $D = 5$ mod 4, with $m = 2^{[D/2]}$. The possible connection between the $D = 11$ case and M-Theory is also discussed.

1 Introduction

A good part of the results presented in this lecture were also discussed in [1] and also presented at the January '98 meeting in Bariloche [2] –where the detailed construction of the superalgebra can be found–, but it was at the meeting covered by these proceedings where these results were first presented.

Three of the four fundamental forces of nature are consistently described by Yang-Mills (**YM**) quantum theories. Gravity, the fourth fundamental interaction, resists quantization in spite of several decades of intensive research in this direction. This is intriguing in view of the fact that General Relativity (**GR**) and YM theories have a deep geometrical foundation: the gauge principle. How come two theories constructed on almost the same mathematical basis produce such radically different physical behaviours? What is the obstruction for the application of the methods of YM quantum field theory to gravity? The final answer to these questions is beyond the scope of this paper, however one can note a difference between YM and GR which might turn

*JOHN SIMON GUGGENHEIM FELLOW

out to be an important clue: YM theory is defined on a fiber bundle, with the connection as the dynamical object, whereas the dynamical fields of GR cannot be interpreted as components of a connection. Therefore, gravitation does not lend itself naturally for a fiber bundle interpretation.

The closest one could get to a connection formulation for GR is the Palatini formalism, with the Hilbert action

$$I[\omega, e] = \int \epsilon_{abcd} R^{ab} \wedge e^a \wedge e^b, \tag{1}$$

where $R^{ab} = d\omega^{ab} + \omega^a_c \wedge \omega^c_b$ is the curvature two-form, and e^a is a local orthonormal frame. This action is sometimes claimed to describe a gauge theory for local translations. However, in our view this is a mistake. If ω and e were the components of the Poincaré connection, under local translations they should transform as

$$\delta\omega^{ab} = 0, \quad \delta e^a = D\lambda^a = d\lambda^a + \omega^a_b \wedge \lambda^b. \tag{2}$$

Invariance of (1) under (2)would require the torsion-free condition,

$$T^a = de^a + \omega^a_b \wedge e^b = 0. \tag{3}$$

This condition is an equation of motion for the action (1). This means that the invariance of the action (1) under (2) could not result from the transformation properties of the fields alone, but it would be a property of their dynamics as well. The torsion-free condition, being one of the field equations, implies that local translational invariance is at best an *on-shell* symmetry, which would probably not survive quantization.

The contradiction stems from the identification between local translations in the base manifold (diffeomorphisms)

$$x^\mu \to x'^\mu = x^\mu + \zeta^\mu(x), \tag{4}$$

–which is a genuine invariance of the action (1)–, and local translations in the tangent space (2).

Since the invariance of the Hilbert action under general coordinate transformations (4) is reflected in the closure of the first-class hamiltonian constraints in the Dirac formalism, one could try to push the analogy between the Hamiltonian constraints H_μ and the generators of a gauge algebra. However, the fact that the constraint algebra requires structure *functions*, which depend on the dynamical fields, is another indication that the generators of diffeomorphism invariance of the theory do not form a Lie algebra but an open algebra (see, e. g., [3]).

More precisely, the subalgebra of spatial diffeomorphisms *is* a genuine Lie algebra in the sense that its structure constants are independent of the dynamical fields of gravitation,

$$[H_i, H'_j] \sim H'_j \delta_{|i} - H'_i \delta_{|j}. \tag{5}$$

In contrast, the generators of timelike diffeomorphisms form an open algebra,

$$[H_\perp, H'_\perp] \sim g^{ij} H'_j \delta_{|i}. \tag{6}$$

This comment is particularly relevant in a CHern-Simons theory, where spatial diffeomorphisms are always part of the true gauge symmetries of the theory. The generators of timelike displacements ($H_\perp$), on the other hand, are combinations of the internal gauge generators and the generators of spatial diffeomorphism, and therefore do not generate independent symmetries [4].

Higher D The minimal requirements for a consistent theory which includes gravity in any dimension are: general covariance and second order field equations for the metric. For $D > 4$ the most general action for gravity satisfying this criterion is a polynomial of degree $[D/2]$ in the curvature, first discussed by Lanczos for $D = 5$ [5] and, in general, by Lovelock [6,7].

First order theory

If the theory contains spinors that couple to gravity, it is necessary to decouple the affine and metric properties of spacetime. A metric formulation is sufficient for spinless point particles and fields because they only couple to the symmetric part of the affine connection, while a spinning particle can "feel" the torsion of spacetime. Thus, it is reasonable to look for a formulation of gravity in which the spin connection (ω_μ^{ab}) and the vielbein (e_μ^a) are dynamically independent fields, with curvature and torsion standing on a similar footing. Thus, the most general gravitational Lagrangian would be of the general form $L = L(\omega, e)$ [8].

Allowing an independent spin connection in four dimensions does not modify the standard picture in practice because any occurrence of torsion in the action leaves the classical dynamics essentially intact. In higher dimensions, however, theories that include torsion can be dynamically quite different from their torsion-free counterparts.

As we shall see below, the dynamical independence of ω^{ab} and e^a also allows defining these gravitation theories in $2n+1$ dimensions on a fiber bundle structure as a Yang-Mills theory, a feature that is not shared by General Relativity except in three dimensions.

2 Supergravity

For some time it was hoped that the nonrenormalizability of GR could be cured by supersymmetry. However, the initial glamour of supergravity (**SUGRA**) as a mechanism for taming the wild ultraviolet divergences of pure gravity, was eventually spoiled by the realization that it too would lead to a nonrenormalizable answer [9]. Again, one can see that SUGRA is not a gauge theory either in the sense of a fiber bundle, and that the local symmetry algebra closes naturally only on shell. The algebra can be made to close off shell at the cost of introducing auxiliary fields, but they are not guaranteed to exist for all D and N [10].

Whether the lack of fiber bundle structure is the ultimate reason for the nonrenormalizability of gravity remains to be proven. However, it is certainly true that if GR could be formulated as a gauge theory, the chances for proving its renormalizability would clearly grow.

In three spacetime dimensions both GR and SUGRA define renormalizable quantum theories. It is strongly suggestive that precisely in 2+1 dimensions both theories can also be formulated as gauge theories on a fiber bundle [11]. It might seem that the exact solvability miracle was due to the absence of propagating degrees of freedom in three-dimensional gravity, but the power counting renormalizability argument rests on the fiber bundle structure of the Chern-Simons form of those systems.

There are other known examples of gravitation theories in odd dimensions which are genuine (off-shell) gauge theories for the anti-de Sitter (**AdS**) or Poincaré groups [12,13,14,15]. These theories, as well as their supersymmetric extensions have propagating degrees of freedom [4] and are CS systems for the corresponding groups as shown in [16].

2.1 From Rigid Supersymmetry to Supergravity

Rigid SUSY can be understood as an extension of the Poincaré algebra by including supercharges which are the "square roots" of the generators of rigid translations, $\{\bar{Q}, Q\} \sim \Gamma \cdot \mathrm{P}$. The basic strategy to generalize this idea to local SUSY was to substitute the momentum $\mathrm{P}_\mu = i\partial_\mu$ by the generators of diffeomorphisms, $\mathcal{H}$, and relate them to the supercharges by $\{\bar{Q}, Q\} \sim \Gamma \cdot \mathcal{H}$. The resulting theory has on-shell local supersymmetry algebra [17].

An alternative point of view –which is the one we advocate here– would be to construct the supersymmetry on the tangent space and not on the base manifold. This approach is more natural if one recalls that spinors provide a basis of irreducible representations for $SO(N)$, and not for $GL(N)$. Thus, spinors are naturally defined relative to a local frame on the tangent space rather than in the coordinate basis. The basic point is to reproduce the 2+1 "miracle" in higher dimensions. This idea has been successfully applied by Chamseddine in five dimensions[13], and by us for pure gravity [14,15] and in supergravity [1,16]. The SUGRA construction has been carried out for spacetimes whose tangent space has AdS symmetry [1], and for its Poincaré contraction in [16].

In [16], a family of theories in odd dimensions, invariant under the supertranslation algebra whose bosonic sector contains the Poincaré generators was presented. The anticommutator of the supersymmetry generators gives a translation plus a tensor "central" extension,

$$\{Q^\alpha, \bar{Q}_\beta\} = -i(\Gamma^a)^\alpha_\beta P_a - i(\Gamma^{abcde})^\alpha_\beta Z_{abcde}, \tag{7}$$

The commutators of $Q, \bar{Q}$ and Z with the Lorentz generators can be read off from their tensorial character. All the remaining commutators vanish. This algebra is the continuation to all odd-dimensional spacetimes of the $D = 10$ superalgebra of van Holten and Van Proeyen [18], and yields supersymmetric theories with off-shell Poincaré superalgebra. The existence of these theories suggests that there should be similar supergravities based on the AdS algebra.

2.2 Assumptions of Standard Supergravity

Three implicit assumptions are usually made in the construction of standard SUGRA:

(i) The fermionic and bosonic fields in the Lagrangian should come in combinations such that their propagating degrees of freedom are equal in number. This is usually achieved by adding to the graviton and the gravitini a number of lower spin fields ($s < 3/2$)[17]. This matching, however, is not necessarily true in AdS space, nor in Minkowski space if a different representation of the Poincaré group (e.g., the adjoint representation) is used [19].

The other two assumptions concern the purely gravitational sector. They are as old as General Relativity itself and are dictated by economy: **(ii)** gravitons are described by the Hilbert action (plus a possible cosmological constant), and, **(iii)** the spin connection and the vielbein are not independent fields but are related through the torsion equation. The fact that the supergravity generators do not form a closed off-shell algebra can be traced back to these asumptions.

The procedure behind **(i)** is tightly linked to the idea that the fields should be in a *vector* representation of the Poincaré group [19] and that the kinetic terms and couplings are such that the counting of degrees of freedom works like in a minimally coupled gauge theory. This assumption comes from the interpretation of supersymmetric states as represented by the in- and out- plane waves in an asymptotically free, weakly interacting theory in a minkowskian background. These conditions are not necessarily met by a CS theory in an asymptotically AdS background. Apart from the difference in background, which requires a careful treatment of the unitary irreducible representations of the asymptotic symmetries [20], the counting of degrees of freedom in CS theories is completely different from the one for the same connection one-forms in a YM theory.

3 Lanczos–Lovelock Gravity

3.1 Lagrangian

For $D > 4$, assumption **(ii)** is an unnecessary restriction on the available theories of gravitation. In fact, as mentioned above, the most general action for gravity –generally covariant and with second order field equations for the metric– is the Lanczos-Lovelock Lagrangian (**LL**). The LL Lagrangian in a D-dimensional Riemannian manifold can be defined in at least four ways:

(**a**) As the most general invariant constructed from the metric and curvature leading to second order field equations for the metric [5,6,7].

(**b**) As the most general D-form invariant under local Lorentz transformations, constructed with the vielbein, the spin connection, and their exterior derivatives, without using the Hogde dual $(*)$ [21].

(**c**) As a linear combination of the dimensional continuation of all the Euler classes of dimension $2p < D$.[7,22]

(**d**) As the most general low energy effective gravitational theory that can be obtained from string theory [23].

Definition (**a**) was historically the first. It is appropriate for the metric formulation and assumes vanishing torsion. Definition (**b**) is slightly more general than the first and allows for a coordinate-independent first-order formulation, and even allows torsion-dependent terms in the action [8]. As a consequence of (**b**), the field configurations that extremize the action obey first order equations for ω and e. Assertion (**c**) gives directly the Lanczos–Lovelock solution as a polynomial of degree $[D/2]$ in the curvature of the form

$$I_G = \int \sum_{p=0}^{[D/2]} \alpha_p L^p, \tag{8}$$

where α_p are arbitrary constants and[a]

$$L_G^p = \epsilon_{a_1 \cdots a_D} R^{a_1 a_2} \cdots R^{a_{2p-1} a_{2p}} e^{a_{2p+1}} \cdots e^{a_D}, \tag{9}$$

where wedge product of forms is understood throughout.

Statement (**d**) reflects the empirical observation that the vanishing of the superstring β-function in $D = 10$ gives rise to an effective Lagrangian of the form (9) [23].

In even dimensions, the last term in the sum is the Euler character, which does not contribute to the equations of motion. However, in the quantum theory, this term in the partition function would assign different weights to nonhomeomorphic geometries.

The large number of dimensionful constants α_p in the LL theory contrasts with the two constants of the EH action (G and Λ) [24,25,14]. This feature could be seem as an indication that renormalizability would be even more remote for the LL theory than in ordinary gravity. However, this is not necessarily so. There are some very special choices of α_p such that the theory becomes invariant under a larger gauge group in odd spacetime dimensions, which could actually improve renormalizability [11,15].

3.2 Equations

Consider the Lovelock action (8), viewed as a functional of the spin connection and the vielbein,

$$I_{LL} = I_{LL}\left[\omega^{ab}, e^a\right]. \tag{10}$$

Varying with respect to the vielbein, the generalized Einstein equations are obtained,

$$\sum_{p=0}^{n-1} \alpha_p (D-2p) \epsilon_{a_1 \cdots a_D} R^{a_1 a_2} \cdots R^{a_{2p-1} a_{2p}} \times e^{a_{2p+1}} \cdots e^{a_{D-1}} = 0. \tag{11}$$

Varying with respect to the spin connection, the torsion equations are found,

$$\sum_{p=0}^{n-1} \alpha_p p (D-2p) \epsilon_{a b a_3 \cdots a_D} R^{a_3 a_4} \cdots R^{a_{2p-1} a_{2p}} \times e^{a_{2p+1}} \cdots e^{a_{D-1}} T^{a_D} = 0. \tag{12}$$

[a] For even and odd dimensions the same expression (9) can be used, but for odd D, Chern-Simons forms for the Lorentz connection could also be included (this point is discussed below).

The presence of the arbitrary coefficients α_p in the action implies that static, spherically symmetric Schwarzschild-like solutions possess a large number of horizons [26], and time-dependent solutions have an unpredictable evolution [22,27]. However, as shown below, for a particular choice of the constants α_p the dynamics is significantly better behaved.

Additional terms containing torsion explicitly can be included in the action. It can be shown, however, that the presence of torsional terms in the Lagrangian does not change the degrees of freedom of gravity in four dimensions. Indeed, the matter-free theory with torsion terms is indistinguishable (at least classically) from GR, [28]. However, in higher dimensions, the situation is completely different [8].

3.3 The vanishing of Classical Torsion

Obviously $T^a = 0$ solves (12). However, for $D > 4$ this equation does not imply vanishing torsion in general. In fact, there are choices of the coefficients α_p and configurations of ω^{ab}, e^a such that T^a is completely arbitrary. On the other hand, as already mentioned, the torsion-free postulate is at best a good description of the classical dynamics only. Thus, an off-shell treatment of gravity should allow for dynamical torsion even in four dimensions. In the first order formulation, the theory has second class constraints due to the presence of a large number of "coordinates" which are actually "momenta" [29], thus complicating the dynamical analysis of the theory.

On the other hand, if torsion is assumed to vanish, ω could be solved as a function of e^{-1} and its first derivatives, but this would restrict the validity of the approach to nonsingular configurations for which $\det(e^a_\mu) \neq 0$. In this framework, the theory has no second class constraints and the number of degrees of freedom is the same as in the Einstein-Hilbert theory, namely $\frac{D(D-3)}{2}$ [22].

3.4 Dynamics and Degrees of Freedom

Imposing $T^a = 0$ from the start, the action is $I{=}I_{LL}[e^a, \omega(e)]$ and varying respect to e, the "1.5 order formalism" [17] is obtained,

$$\delta I {=} \frac{\delta I_{LL}}{\delta e^a}\delta e^a + \frac{\delta I_{LL}}{\delta \omega^{bc}}\frac{\delta \omega^{bc}}{\delta e^a}\delta e^a. \tag{13}$$

Assuming $\frac{\delta I_{LL}}{\delta \omega^{bc}} = 0$ the equations of motion consist of the Einstein equations (11), defined on a restricted configuration space.

For $D \leq 4$, $T^a = 0$ is the unique solution of eqn.(12). In those dimensions, the different variational principles (first-, second- and 1.5-th order) are classically equivalent in the absence of sources. On the contrary, for $D > 4$, $T^a = 0$ is not logically necessary and is therefore unjustified.

The LL–Lagrangians (9) include the Einstein-Hilbert (**EH**) theory as a particular case, but they are dynamically very different in general. The classical solutions of the LL theory are not perturbatively related to those of the Einstein theory. For instance, it was observed that the time evolution of the classical solutions in the LL theory starting from a generic initial state can be unpredictable, whereas the EH theory defines a well-posed Cauchy problem.

It can also be seen that even for some simple minisuperspace models, the dynamics could become quite messy because the equations of motion are not deterministic in the classical sense, due to the vanishing of some eigenvalues of the Hessian matrix on critical surfaces in phase space [22,27].

3.5 Choice of Coefficients

At least for some simple minisuperspace geometries the indeterminate classical evolution can be avoided if the coefficients are chosen so that the Lagrangian is based on the connection for the AdS group,

$$\alpha_p l^{D-2p} = \begin{cases} (D-2p)^{-1}\binom{n-1}{p}, & D = 2n-1 \\ \binom{n}{p}, & D = 2n. \end{cases} \tag{14}$$

This corresponds to the Born-Infeld theory in even dimensions [25], and to the AdS Chern-Simons theory in odd dimensions [14,12,13].

3.5.1 $D = 2n$: Born-Infeld Gravity

In even dimensions the choice (14) gives rise to a Lagrangian of the form

$$L = \kappa\epsilon_{a_1\cdots a_D}(R^{a_1a_2} + \frac{e^{a_1}e^{a_2}}{l^2})\cdots(R^{a_{D-1}a_D} + \frac{e^{a_{D-1}}e^{a_D}}{l^2}). \tag{15}$$

This is the Pfaffian of the two–form $R^{ab} + \frac{1}{l^2}e^a e^b$, and, in this sense it can be written in the Born-Infeld-like form,

$$L = \kappa\sqrt{\det(R^{ab} + \frac{1}{l^2}e^a e^b)}. \tag{16}$$

The combinations $R^{ab} + \frac{1}{l^2}e^a e^b$ are the components of the AdS curvature (c.f.(19) below). This seems to suggest that the system might be naturally described in terms of an AdS connection [30]. However, this is not the case: In even dimensions, the Lagrangian (15) is invariant under local Lorentz transformations and not under the entire AdS group. As will be shown below, it is possible, in odd dimensions, to construct gauge invariant theories of gravity under the full AdS group.

3.5.2 $D = 2n-1$: AdS Gauge Gravity

The odd-dimensional case was discussed in [12,13], and later also in [14]. Consider the action (8) with the choice given by (14) for $D = 2n-1$. The constant parameter l has dimensions of length and its purpose is to render the action dimensionless. This also allows the interpretation of ω and e as components of the AdS connection [25], $A = \frac{1}{2}\omega^{ab}J_{ab} + e^a J_{aD+1} = \frac{1}{2}W^{AB}J_{AB}$, where

$$W^{AB} = \begin{bmatrix} \omega^{ab} & e^a/l \\ -e^b/l & 0 \end{bmatrix}, \quad A, B = 1, ...D+1. \tag{17}$$

The resulting Lagrangian is the Euler-CS form. Its exterior derivative is the Euler form in $2n$ dimensions,

$$\begin{aligned} dL^{AdS}_{G\,2n-1} &= \kappa\epsilon_{A_1\cdots A_{2n}} R^{A_1A_2}\cdots R^{A_{2n-1}A_{2n}} \\ &= \kappa\mathcal{E}_{2n}, \end{aligned} \tag{18}$$

where $R^{AB} = dW^{AB} + W^A_C W^{CB}$ is the AdS curvature, which contains the Riemann and torsion tensors,

$$R^{AB} = \begin{bmatrix} R^{ab} + \frac{1}{l^2}e^a e^b & T^a/l \\ -T^b/l & 0 \end{bmatrix}. \tag{19}$$

The constant κ is quantized [15] (in the following we will set $\kappa = l = 1$).

In general, a Chern-Simons Lagrangian in $2n-1$ dimensions is defined by the condition that its exterior derivative be an invariant homogeneous polynomial of degree n in the curvature, that is, a characteristic class. In the case above, (19) defines the CS form for the Euler class $2n$-form.

A generic CS Lagrangian in $2n-1$ dimensions for a Lie algebra g can be defined by

$$dL^g_{2n-1} = \langle \mathbf{F}^n \rangle, \tag{20}$$

where $\langle\ \rangle$ stands for a multilinear function in the Lie algebra g, invariant under cyclic permutations such as Tr, for an ordinary Lie algebra, or STr, in the case of a superalgebra. In the case above, the only nonvanishing brackets in the algebra are

$$\langle J_{A_1A_2}, \cdots, J_{A_{D-1}A_D} \rangle = \epsilon_{A_1\cdots A_D}. \tag{21}$$

3.5.3 $D = 2n - 1$: *Poincaré Gauge Gravity*

Starting from the AdS theory (19) in odd dimensions, a Wigner- Inönü contraction deforms the AdS algebra into the Poincaré one. The same result is also obtained choosing $\alpha_p = \delta_p^n$. Then, the Lagrangian (8) becomes:

$$L_G^P = \epsilon_{a_1 \cdots a_D} R^{a_1 a_2} \cdots R^{a_{D-2} a_{D-1}} e^{a_D}. \tag{22}$$

In this way the local symmetry group of (8) is extended from Lorentz ($SO(D-1,1)$) to Poincaré ($ISO(D-1,1)$). Analogously to the anti-de Sitter case, one can see that the action depends on the Poincaré connection: $\mathbf{A} = e^a P_a + \frac{1}{2}\omega^{ab} J_{ab}$. It is straightforward to verify the invariance of the action under local translations,

$$\delta e^a = D\lambda^a, \quad \delta\omega^{ab} = 0, \tag{23}$$

Here D stands for covariant derivative in the Lorentz connection. If λ is the Lie algebra-valued zero-form, $\lambda = \lambda^a P_a$, the transformations (23) are read from the general gauge transformation for the connection, $\delta\mathbf{A} = \nabla\lambda$, where ∇ is the covariant derivative in the Poincaré connection.

Moreover, the Lagrangian (22) is a Chern-Simons form. Indeed, with the curvature for the Poincaré algebra, $\mathbf{F} = d\mathbf{A} + \mathbf{A} \wedge \mathbf{A} = \frac{1}{2} R^{ab} J_{ab} + T^a P_a$, L_G^P satisfies

$$dL_G^P = \langle \mathbf{F}^{n+1} \rangle, \tag{24}$$

where the only nonvanishing components in the bracket are

$$\langle J_{a_1 a_2}, \cdots, J_{a_{D-2} a_{D-1}}, P_{a_D} \rangle = \epsilon_{a_1 \cdots a_D}. \tag{25}$$

Thus, the Chern character for the Poincaré group is written in terms of the Riemman curvature and the torsion as

$$\langle \mathbf{F}^3 \rangle = \epsilon_{a_1 \cdots a_D} R^{a_1 a_2} \cdots R^{a_{D-2} a_{D-1}} T^{a_D}. \tag{26}$$

The simplest example of this is ordinary gravity in 2+1 dimensions, where the Einstein-Hilbert action with cosmological constant is a genuine *gauge* theory of the AdS group, while for zero cosmological constant it is invariant under *local* Poincaré transformations. Although this gauge invariance of 2+1 gravity is not always emphasized, it lies at the heart of the proof of integrability of the theory [11].

4 AdS Gauge Gravity

As shown above, the LL action assumes spacetime to be a Riemannian, torsion-free, manifold. That assumption is justified *a posteriori* by the observation that $T^a = 0$ is always a solution of the classical equations, and means that e and ω are not dynamically independent. This is the essence of the second order or metric approach to GR, in which distance and parallel transport are not independent notions, but are related through the Christoffel symbol. There is no fundamental justification for this assumption and this was the issue of the historic discussion between Einstein and Cartan [31].

In four dimensions, the equation $T^a = 0$ is algebraic and could in principle be solved for ω in terms of the remaining fields. However, for $D > 5$, CS gravity has more degrees of freedom than those encountered in the corresponding second order formulation [4]. This means that the CS gravity action has propagating degrees of freedom for the spin connection. This is a compelling argument to seriously consider the possibility of introducing torsion terms in the Lagrangian from the start.

Another consequence of imposing a dynamical dependence between ω and e through the torsion-free condition is that it spoils the possibility of interpreting the local translational invariance as a gauge symmetry of the action. Consider the action of the Poincaré group on the fields as given by (23); taking $T^a \equiv 0$ implies

$$\delta\omega^{ab} = \frac{\delta\omega^{ab}}{\delta e^c}\delta e^c \neq 0, \tag{27}$$

which would be inconsistent with the transformation of the fields under local translations (2). Thus, the spin connection and the vielbein –the soldering between the base manifold and the tangent space– cannot be identified as the compensating fields for local Lorentz rotations and translations, respectively.

In our construction ω and e are assumed to be dynamically independent and thus torsion necessarily contains propagating degrees of freedom, represented by the contorsion tensor $k_\mu^{ab} := \omega_\mu^{ab} - \bar{\omega}_\mu^{ab}(e,...)$, where $\bar{\omega}$ is the spin connection which solves the (algebraic) torsion equation in terms of the remaining fields.

The generalization of the Lovelock theory to include torsion explicitly can be obtained assuming definition **(b)**. This is a cumbersome problem due to the lack of a simple algorithm to classify all possible invariants constructed from e^a, R^{ab} and T^a. In Ref. [8] a useful "recipe" to generate all those invariants is given.

4.1 The Two Families of AdS Theories

Similarly to the theory discussed in section III, the torsional additions to the Lagrangian bring in a number of arbitrary dimensionful coefficients β_k, analogous to the α_p's. Also in this case, one can try choosing the β's in such a way as to enlarge the local Lorentz invariance into an AdS gauge symmetry. If no additional structure (e.g., inverse metric, Hodge-$*$, etc.) is assumed, AdS invariants can only be produced in dimensions $4k$ and $4k-1$.

The proof of this claim is as follows: invariance under AdS requires that the D-form be at least Lorentz invariant. Then, in order for these scalars to be invariant under AdS as well, it is necessary and sufficient that they be expressible in terms of the AdS connection (17). As is well-known (see, e.g., [32]), in even dimensions, the only D-form invariant under $SO(N)$ constructed according to the recipe mentioned above are[b] the Euler character (for $N = D$), and the Chern characters (for any N). Thus, the only AdS invariant D-forms are the Euler class, and linear conbinations of products of the type

$$P_{r_1 \cdots r_s} = c_{r_1} \cdots c_{r_s}, \tag{28}$$

with $2(r_1 + r_2 + \cdots + r_s) = D$, where

$$c_r = \mathrm{Tr}(\mathbf{F}^r), \tag{29}$$

defines the r-th Chern character of $SO(N)$. Now, since the curvature two-form $\mathbf{F}$ in the vectorial representation is antisymmetric in its indices, the exponents $\{r_j\}$ are necessarily even, and therefore (28) vanishes unless D is a multiple of four. Thus, one arrives at the following lemmas:

Lemma: 1 For $D = 4k$, the only D-forms built from e^a, R^{ab} and T^a, invariant under the AdS group, are the Chern characters for $SO(D+1)$.

Lemma: 2 For $D = 4k+2$, there are no AdS-invariant D-forms constructed from e^a, R^{ab} and T^a.

In view of this, it is clear why attempts to construct gravitation theories with local AdS invariance in even dimensions have been unsuccessful [30,33].

Since the forms $P_{r_1 \cdots r_s}$ are closed, they are at best boundary terms in $4k$ dimensions –which do not contribute to the classical equations, but could assign different weights to configurations with nontrivial torsion in the quantum theory. In other words, they can be locally expressed as

$$P_{r_1 \cdots r_s} = dL^{AdS}_{\{r\}4k-1}(W). \tag{30}$$

[b]For simplicity we will not always distinguish between different signatures. Thus, if no confussion can occur, the AdS group in D dimensions will also be denoted as $SO(D+1)$. The de Sitter case can be obtained replacing α_p by $(-1)^p \alpha_p$ in (14).

Thus, for each collection $\{r\}$, the $(4k-1)$- form $L^{AdS}_{\{r\}4k-1}$ defines a Lagrangian for the AdS group in $4k-1$ dimensions. It takes direct computation to see that these Lagrangians involve torsion explicitly. These results are summarized in the following

Theorem: There are two families of gravitational first order Lagrangians for e and ω, invariant under local AdS transformations:

a: Euler-Chern-Simons form in $D = 2n-1$, whose exterior derivative is the Euler character in dimension $2n$, which do not involve torsion explicitly, and

b: Pontryagin-Chern-Simons forms in $D = 4k-1$, whose exterior derivatives are the Chern characters in $4k$ dimensions, which involves torsion explicitly.

It must be stressed that locally AdS-invariant gravity theories only exist in odd dimensions. They are *genuine* gauge systems, whose action comes from topological invariants in one dimension above. These topological invariants can be written as the trace of a homogeneous polynomial of degree n in the AdS curvature. Obviously, for dimensions $4k-1$ both **a**- and **b**-families exist. The most general Lagrangian of this sort is a linear combination of the two families.

An important difference between these two families is that under a parity transformation the first is even while the second is odd [c]. The parity invariant family has been extensively studied in [12,13,14,25]. In what follows we concentrate on the construction of the pure gravity sector as a gauge theory which is parity-odd. This construction was discussed in [34], and also briefly in [1,2].

4.2 Even dimensions

In $D = 4$, the the only local Lorentz-invariant 4-forms constructed with the recipe just described are [8]:

$$\begin{aligned}
\mathcal{E}_4 &= \epsilon_{abcd} R^{ab} R^{cd} \\
L_{EH} &= \epsilon_{abcd} R^{ab} e^c e^d \\
L_C &= \epsilon_{abcd} e^a e^b e^c e^d \\
C_2 &= R^{ab} R_{cd} \\
L_{T_1} &= R^{ab} e_a e_b
\end{aligned}$$

[c] Parity is understood here as an inversion of one coordinate, both in the tangent space and in the base manifold. Thus, for instance the Euler character is invariant under parity, while the Lorentz Chern characters and the torsional terms are parity violating.

$$L_{T_2} = T^a T_a.$$

The first three terms are even under parity and the rest are odd. Of these, $\mathcal{E}_4$ and C_2 are topological invariant densities (closed forms): the Euler character and the second Chern character for $SO(4)$,respectively. The remaining four terms define the most general gravity action in four dimensions,

$$I = \int_{M_4} [\alpha L_{EH} + \beta L_C + \gamma L_{T1} + \rho L_{T2}] . \tag{31}$$

It can also be seen, that by choosing $\gamma = -\rho$, the last two terms are combined into a topological invariant density (the Nieh-Yan form). Thus, with this choice the odd part of the action becomes a boundary term. Furthermore, C_2, L_{T_1} and L_{T_2} can be combined into the second Chern character of the AdS group,

$$R^a{}_b R^b{}_a + 2(T^a T_a - 2R^{ab} e_a e_b) = R^A{}_B R^B{}_A. \tag{32}$$

This is the only AdS invariant constructed with e^a, ω^{ab} and their exterior derivatives alone, confirming that there are no locally AdS invariant gravities in four dimensions.

In general, the only AdS-invariant functionals in higher dimensions can be written in terms of the AdS curvature as [8]

$$\tilde{I}_{r_1 \cdots r_s} = \int_M C_{r_1} \cdots C_{r_s}, \tag{33}$$

or linear combinations thereof, where $C_r = Tr[(R^A_B)^r]$ is the r-th Chern character for the AdS group. For example, en $D = 8$ the Chern characters for the AdS group are

$$\begin{aligned} Tr[(R^A_B)^4] &= C_4, \\ Tr[(R^A_B)^2]_\wedge Tr[(R^A_B)^2] &= (C_2)^2. \end{aligned} \tag{34}$$

Similar Chern classes are also found for $D = 4k$. (As already mentioned, $\tilde{I}_{r_1 \cdots r_s}$ vanishes if one of the r's is odd, which is the case in $4k+2$ dimensions.)

Thus, there are no AdS-invariant gauge theories in even dimensions.

4.3 Odd dimensions

The simplest example is found in three spacetime dimensions where there are two locally AdS-invariant Lagrangians, namely, the Einstein-Hilbert with cosmological constant,

$$L_{G\,3}^{AdS} = \epsilon_{abc}[R^{ab}e^c + \frac{1}{3l^2}e^a e^b e^c], \tag{35}$$

and the "exotic" Lagrangian

$$L_{T\,3}^{AdS} = L_3^*(\omega) + 2e_a T^a, \tag{36}$$

where

$$L_3^* \equiv \omega_b^a d\omega_a^b + \frac{2}{3}\omega_b^a \omega_c^b \omega_a^c, \tag{37}$$

is the Lorentz Chern-Simons form. Note that in (36), the local AdS symmetry fixes the relative coefficient between $L_3^*(\omega)$, and the torsion term $e_a T^a$. The most general action for gravitation in $D = 3$, which is invariant under $SO(4)$ is therefore a linear combination $\alpha L_{G\,3}^{AdS} + \beta L_{T\,3}^{AdS}$.

For $D = 4k - 1$, the number of possible exotic forms grows as the partitions of k, in correspondence with the number of composite Chern invariants of the form $P_{\{r\}} = \prod_j C_{r_j}$. The most general Lagrangian in $4k-1$ dimensions takes the form $\alpha L_{G\,4k-1}^{AdS} + \beta_{\{r\}} L_{T\,\{r\}\,4k-1}^{AdS}$, where $dL_{T\,\{r\}\,4k-1}^{AdS} = P_{\{r\}}$, with $\sum_j r_j = 4k$. These Lagrangians have proper dynamics and, unlike the even dimensional cases, they are not boundary terms. For example, in seven dimensions one finds [34,35]

$$\begin{aligned} L_{T\,7}^{AdS} &= \beta_{2,2}[R^a{}_b R^b{}_a + 2(T^a T_a - R^{ab} e_a e_b)] L_{T\,3}^{AdS} \\ &+ \beta_4 [L_7^*(\omega) + 2(T^a T_a + R^{ab} e_a e_b) T^a e_a + 4 T_a R^a{}_b R^b{}_c e^c], \end{aligned}$$

where L_{2n-1}^* is the Lorentz-CS (2n-1)-form,

$$dL_{2n-1}^*(\omega) = Tr[(R_b^a)^n]. \tag{38}$$

Summarizing: The requirement of local AdS symmetry is rather strong and has the following consequences:

- Locally AdS invariant theories of gravity exist in odd dimensions only.
- For $D = 4k - 1$ there are two families: one involving only the curvature and the vielbein (Euler Chern-Simons form), and the other involving torsion explicitly in the Lagrangian. These families are even and odd under space reflections, respectively.

- For $D = 4k+1$ only the Euler-Chern-Simons forms exist. These ar parity even and don't involve torsion explicitly.

5 Exact Solutions

As stressed here, the local symmetry of odd-dimensional gravity can be extended from Lorentz to AdS by an appropriate choice of the free coefficients in the action. The resulting Lagrangians (with or without torsion terms), are Chern-Simons D-forms defined in terms of the AdS connection $\mathbf{A}$, whose components include the vielbein and the spin connection [see eqn. (17)]. This implies that the field equations (11,12) obtained by varying the vielbein and the spin connection respectively, can be written in an AdS-covariant form

$$< \mathbf{F}^{n-1} J_{AB} >= 0, \tag{39}$$

where $\mathbf{F}= \frac{1}{2} R^{AB} J_{AB}$ is the AdS curvature with R^{AB} given by (19) and J_{AB} are the AdS generators.

It is easily checked that any locally AdS spacetime is a solution of (39). Apart from anti-de Sitter space itself, some interesting spacetimes with this feature are the topological black holes of Ref. [36], and some "black branes" with constant curvature worldsheet [37]. For any D, there is also a unique static, spherically symmetric, asymptotically AdS black hole solution [14], as well as their topological extensions which have nontrivial event horizons [38].

Exact solutions of the form $AdS_4 \times S^{D-4}$ have also been found [39] [d] as well as alternative four-dimensional cosmological models.

All of the above geometries can be extended into solutions of the gravitational Born-Infeld theory (16) in even dimensions. Friedmann-Robertson-Walker like cosmologies have been shown to exist in even dimensions [25], and it could be expected that similar solutions exists in odd dimensions as well.

6 Chern-Simons Supergravities

We now consider the supersymmetric extensions of the locally AdS theories defined above. The idea is to enlarge the AdS algebra incorporating SUSY generators. The closure of the algebra (Jacobi identity) forces the addition of further bososnic generators as well [18]. In order to accomodate spinors in a

[d] The de-Sitter case ($\Lambda > 0$) was discussed in [40] for the torsion-free theory. Changing the sign in the cosmological constant has deep consequences. In fact, the solutions are radically different, and locally supersymmetric extensions for positive cosmological constant don't exist in general.

natural way, it is useful to cast the AdS generators in the spinor representation of $SO(D+1)$. In particular, one can write,

$$dL_{T\,4k-1}^{AdS} = \frac{-1}{2^{4k}} Tr[(R^{AB}\Gamma_{AB})^{2k}]. \tag{40}$$

which is a particular form of (20) where $\langle\rangle$ has been replaced by the ordinary trace over spinor indices in this representation.

Other possibilities of the form $\langle \mathbf{F}^{n-p} \rangle \langle \mathbf{F}^{p} \rangle$, are not necessary to reproduce the minimal supersymmetric extensions of AdS containing the Hilbert action. In the supergravity theories discussed below, the gravitational sector is given by $\pm\frac{1}{2^n} L_{G\,2n-1}^{AdS} - \frac{1}{2} L_{T\,2n-1}^{AdS}$. The $\pm$ sign corresponds to the two choices of inequivalent representations of Γ's, which in turn reflect the two chiral representations in $D+1$. As in the three-dimensional case, the supersymmetric extensions of L_G or any of the exotic Lagrangians such as L_T, require using both chiralities, thus doubling the algebras. Here we choose the + sign, which gives the minimal superextension [34].

The bosonic theory (40) is our starting point. The idea now is to construct its supersymmetric extension. For this, we need to express the adjoint representation in terms of the Dirac matrices of the appropriate dimension. This is always possible because the generators of the Dirac algebra, $\{I, \Gamma^a, \Gamma^{ab}, ...\}$, provide a basis for the space of square matrices. The advantage of this approach is that it gives an explicit representation of the algebra and writing the Lagrangians is straightforward.

The supersymmetric extensions of the AdS algebras in $D = 2, 3, 4$, mod 8, were studied by van Holten and Van Proeyen in [18]. They added one Majorana supersymmetry generator to the AdS algebra and found all the $N = 1$ extensions demanding closure of the full superalgebra. In spite of the fact that the algebra for $N = 1$ AdS supergravity in eleven dimensions was conjectured in 1978 to be $osp(32|1)$ by Cremer, Julia and Scherk [41], and this was confirmed in [18], nobody constructed a supergravity action for this algebra in the intervening twenty years.

One reason for the lack of interest in the problem might have been the fact that the $osp(32|1)$ algebra contains generators which are Lorentz tensors of rank higher than two.In the past, supergravity algebras were traditionally limited to generators which are Lorentz tensors up to second rank. This constraint was based on the observation that elementary particle states of spin higher than two would be inconsistent [42]. However, this does not rule out the relevance of those tensor generators in theories of extended objects [43]. In fact, it is quite common nowadays to find algebras like the M–brane superalgebra [44,45],

$$\{Q{\cdot}\bar{Q}\} \sim \Gamma^a P_a + \Gamma^{ab} Z_{ab} + \Gamma^{abcde} Z_{abcde}. \tag{41}$$

6.1 Superalgebra and Connection

The smallest superalgebra containing the AdS algebra in the bosonic sector is found following the same approach as in [18], but lifting the restriction of $N = 1$ [34]. The result, for odd $D > 3$ is (see [2] for details)

D	S-Algebra	Conjugation Matrix	Internal Metric
$8k-1$	$osp(N\|m)$	$C^T = C$	$u^T = -u$
$8k+3$	$osp(m\|N)$	$C^T = -C$	$u^T = u$
$4k+1$	$su(m\|N)$	$C^\dagger = C$	$u^\dagger = u$

In each of these cases, $m = 2^{[D/2]}$ and the connection takes the form

$$\begin{aligned} \mathbf{A} = & \frac{1}{2}\omega^{ab} J_{ab} + e^a J_a + \frac{1}{r!} b^{[r]} Z_{[r]} + \\ & \frac{1}{2}(\bar{\psi}^i Q_i - \bar{Q}^i \psi_i) + \frac{1}{2} a_{ij} M^{ij}. \end{aligned} \tag{42}$$

The generators J_{ab}, J_a span the AdS algebra and the Q^i_α's generate (extended) supersymmetry transformations. The Q's transform in a vector representation under the action of M_{ij} and as spinors under the Lorentz group. Finally, the Z's complete the extension of AdS into the larger algebras $so(m)$, $sp(m)$ or $su(m)$, and $[r]$ denotes a set of r antisymmetrized Lorentz indices.

In (42) $\bar{\psi}^i = \psi^T_j C u^{ji}$ ($\bar{\psi}^i = \psi^\dagger_j C u^{ji}$ for $D = 4k+1$), where C and u are given in the table above. These algebras admit $(m+N) \times (m+N)$ matrix representations [30], where the J and Z have entries in the $m \times m$ block, the M_{ij}'s in the $N \times N$ block, while the fermionic generators Q have entries in the complementary off-diagonal blocks.

Under a gauge transformation, $\mathbf{A}$ transforms by $\delta\mathbf{A} = \nabla\lambda$, where ∇ is the covariant derivative for the same connection $\mathbf{A}$. In particular, under a supersymmetry transformation, $\lambda = \bar{\epsilon}^i Q_i - \bar{Q}^i \epsilon_i$, and

$$\delta_\epsilon \mathbf{A} = \begin{bmatrix} \epsilon^k \bar{\psi}_k - \psi^k \bar{\epsilon}_k & D\epsilon_j \\ -D\bar{\epsilon}^i & \bar{\epsilon}^i \psi_j - \bar{\psi}^i \epsilon_j \end{bmatrix}, \tag{43}$$

where D is the covariant derivative on the bosonic connection, $D\epsilon_j = (d + \frac{1}{2}[e^a \Gamma_a + \frac{1}{2}\omega^{ab}\Gamma_{ab} + \frac{1}{r!} b^{[r]} \Gamma_{[r]}])\epsilon_j - a^i_j \epsilon_i$.

6.2 D=5 Supergravity

In this case, as in every dimension $D = 4k+1$, there is no torsional Lagrangians L_T due to the vanishing of the Pontrjagin $4k+2$-forms for the Riemann cirvature. This fact implies that the local supersymmetric extension will be of the form $L = L_G + \cdots$.

As shown in the previous table, the appropriate AdS superalgebra in five dimensions is $su(2,2|N)$, whose generators are $K, J_a, J_{ab}, Q^\alpha, \bar{Q}_\beta, M^{ij}$, with $a,b = 1,...,5$ and $i,j = 1,...,N$. The connection is $\mathbf{A} = bK + e^a J_a + \frac{1}{2}\omega^{ab}J_{ab} + a_{ij}M^{ij} + \bar{\psi}^i Q_i - \bar{Q}^j \psi_j$, so that in the adjoint representation

$$\mathbf{A} = \begin{bmatrix} \Omega^\alpha_\beta & \psi^\alpha_j \\ -\bar{\psi}^i_\beta & A^i_j \end{bmatrix}, \tag{44}$$

with $\Omega^\alpha_\beta = \frac{1}{2}(\frac{i}{2}bI + e^a\Gamma_a + \omega^{ab}\Gamma_{ab})^\alpha_\beta$, $A^i_j = \frac{i}{N}\delta^i_j b + a^i_j$, and $\bar{\psi}^i_\beta = \psi^{\dagger\alpha j}G_{\alpha\beta}$. Here G is the Dirac conjugate (e. g., $G = i\Gamma^0$). The curvature is

$$\mathbf{F} = \begin{bmatrix} \bar{R}^\alpha_\beta & D\psi^\alpha_j \\ -D\bar{\psi}^i_\beta & \bar{F}^i_j \end{bmatrix} \tag{45}$$

where

$$\begin{aligned} D\psi^\alpha_j &= d\psi^\alpha_j + \Omega^\alpha_\beta\psi^\beta_j - A^i_j\psi^\alpha_i, \\ \bar{R}^\alpha_\beta &= R^\alpha_\beta - \psi^\alpha_i\bar{\psi}^i_\beta, \\ \bar{F}^i_j &= F^i_j - \bar{\psi}^i_\beta\psi^\beta_j. \end{aligned} \tag{46}$$

Here $F^i_j = dA^i_j + A^i_k A^k_j + \frac{i}{N}db\delta^i_j$ is the $su(N)$ curvature, and $R^\alpha_\beta = d\Omega^\alpha_\beta + \Omega^\alpha_\sigma\Omega^\sigma_\beta$ is the $u(2,2)$ curvature. In terms of the standard $(2n-1)$-dimensional fields, R^α_β can be written as

$$R^\alpha_\beta = \frac{i}{4}db\delta^\alpha_\beta + \frac{1}{2}\left[T^a\Gamma_a + (R^{ab} + e^a e^b)\Gamma_{ab}\right]^\alpha_\beta. \tag{47}$$

In six dimensions the only invariant form is

$$P = iStr\left[\mathbf{F}^3\right], \tag{48}$$

which in this case reads

$$\begin{aligned} P = {} & Tr\left[R^3\right] - Tr\left[F^3\right] \\ & + 3\left[D\bar{\psi}(\bar{R} + \bar{F})D\psi - \bar{\psi}(R^2 - F^2 + [R - F](\psi)^2)\psi\right], \end{aligned} \tag{49}$$

where $(\psi)^2 = \bar{\psi}\psi$. The resulting five-dimensional C-S density can de descompossed as a sum a a gravitational part, a b-dependent piece, a $su(N)$ gauge part, and a fermionic term,

$$L = L_G^{AdS} + L_b + L_{su(N)} + L_F, \tag{50}$$

with

$$
\begin{aligned}
L_G^{AdS} &= \tfrac{1}{8}\epsilon_{abcde}(R^{ab}R^{cd}e^e + \tfrac{2}{3}R^{ab}e^ce^de^e + \tfrac{1}{5}e^ae^be^ce^de^e) \\
L_b &= -(\tfrac{1}{N^2} - \tfrac{1}{4^2})(db)^2b + \tfrac{3}{4}(T^aT_a - R^{ab}e_ae_b - \tfrac{1}{2}R^{ab}R_{ab})b \\
&+\tfrac{3}{N}bf^i_jf^j_i \\
L_{su(N)} &= -(a^i_jda^j_kda^k_i + a^i_ja^j_ka^k_lda^l_i + \tfrac{3}{5}a^i_ja^j_ka^k_la^l_ma^m_i) \\
L_F &= \tfrac{3}{2}\left[\bar{\psi}(\bar{R}+\bar{F})D\psi - \tfrac{1}{2}(\psi)^2(\bar{\psi}D\psi)\right].
\end{aligned}
\tag{51}
$$

The action is invariant under local gauge transformations, which contain the local SUSY transformations

$$
\begin{aligned}
\delta e^a &= -\tfrac{1}{2}(\bar{\epsilon}^i\Gamma^a\psi_i - \overline{\psi}^i\Gamma^a\epsilon_i) \\
\delta\omega^{ab} &= \tfrac{1}{4}(\bar{\epsilon}^i\Gamma^{ab}\psi_i - \overline{\psi}^i\Gamma^{ab}\epsilon_i) \\
\delta b &= i(\bar{\epsilon}^i\psi_i - \overline{\psi}^i\epsilon_i \\
\delta\psi_i &= D\epsilon_i \\
\delta\overline{\psi}^i &= D\bar{\epsilon}^i \\
\delta a^i_j &= i(\bar{\epsilon}^i\psi_j - \overline{\psi}^i\epsilon_j).
\end{aligned}
\tag{52}
$$

As in $2+1$ dimensions, the Poincaré supergravity theory is recovered contracting the super *AdS* group. Consider the following rescaling of the fields

$$
\begin{aligned}
e^a &\to \tfrac{1}{\alpha}e^a \\
\omega^{ab} &\to \omega^{ab} \\
b &\to \tfrac{1}{3\alpha}b \\
\psi_i &\to \tfrac{1}{\sqrt{\alpha}}\psi_i \\
\overline{\psi}^i &\to \tfrac{1}{\sqrt{\alpha}}\overline{\psi}^i \\
a^i_j &\to a^i_j.
\end{aligned}
\tag{53}
$$

Then, if the gravitational constant is also rescaled as $\kappa \to \alpha\kappa$, in the limit $\alpha \to \infty$ the action becomes that in [16], plus a $su(N)$ CS form,

$$
I = \frac{1}{8}\int[\epsilon_{abcde}R^{ab}R^{cd}e^e - R^{ab}R_{ab}b - \tag{54}
$$

$$
2R^{ab}(\overline{\psi}^i\Gamma_{ab}D\psi_i + D\overline{\psi}^i\Gamma_{ab}\psi_i) + L_{su(N)}].
$$

The rescaling (53) induces a contraction of the super *AdS* algebra $su(m|N)$ into [super Poincaré]$\otimes su(N)$, where the second factor is an automorfism.

6.3 D=7 Supergravity

The smallest AdS superalgebra in seven dimensions is $osp(2|8)$. The connection (42) is $\mathbf{A} = \frac{1}{2}\omega^{ab}J_{ab} + e^a J_a + \bar{Q}^i\psi_i + \frac{1}{2}a_{ij}M^{ij}$, where M^{ij} are the generators of $sp(2)$. In the representation given above, the bracket $\langle\ \rangle$ is the supertrace and, in terms of the component fields appearing in the connection, the CS form is

$$L_7^{osp(2|8)}(\mathbf{A}) = 2^{-4}L_{G\,7}^{AdS}(\omega,e) - \frac{1}{2}L_{T\,7}^{AdS}(\omega,e) \\ -L_7^{*sp(2)}(a) + L_F(\psi,\omega,e,a). \tag{55}$$

Here the fermionic Lagrangian is

$$\begin{aligned} L_F = {} & 4\bar{\psi}^j(R^2\delta^i_j + Rf^i_j + (f^2)^i_j)D\psi_i \\ & +4(\bar{\psi}^i\psi_j)[(\bar{\psi}^j\psi_k)(\bar{\psi}^k D\psi_i) - \bar{\psi}^j(R\delta^k_i + f^k_i)D\psi_k] \\ & -2(\bar{\psi}^i D\psi_j)[\bar{\psi}^j(R\delta^k_i + f^k_i)\psi_k + D\bar{\psi}^j D\psi_i], \end{aligned}$$

where $f^i_j = da^i_j + a^i_k a^k_j$, and $R = \frac{1}{4}(R^{ab} + e^a e^b)\Gamma_{ab} + \frac{1}{2}T^a\Gamma_a$ are the $sp(2)$ and $so(8)$ curvatures, respectively. The supersymmetry transformations (43) read

$$\delta e^a = \tfrac{1}{2}\bar{\epsilon}^i\Gamma^a\psi_i \qquad \delta\omega^{ab} = -\tfrac{1}{2}\bar{\epsilon}^i\Gamma^{ab}\psi_i$$

$$\delta\psi_i = D\epsilon_i \qquad \delta a^i_j = \bar{\epsilon}^i\psi_j - \bar{\psi}^i\epsilon_j.$$

Standard seven-dimensional supergravity is an $N = 2$ theory (its maximal extension is N=4), whose gravitational sector is given by the Einstein-Hilbert action with cosmological constant and with an $osp(2|8)$ invariant background[46,47]. In the case presented here, the extension to larger N is straighforward: the index i is allowed to run from 2 to $2s$, and the Lagrangian is a CS form for $osp(2s|8)$.

6.4 D=11 Supergravity

In this case, the smallest AdS superalgebra is $osp(32|1)$ and the connection is $\mathbf{A} = \frac{1}{2}\omega^{ab}J_{ab} + e^a J_a + \frac{1}{5!}b^{abcde}J_{abcde} + \bar{Q}\psi$, where b is a totally antisymmetric fifth-rank Lorentz tensor one-form. Now, in terms of the elementary bosonic and fermionic fields, the CS form in (20) reads

$$L_{11}^{osp(32|1)}(\mathbf{A}) = \mathrm{L}_{11}^{sp(32)}(\Omega) + \mathrm{L_F}(\Omega,\psi), \tag{56}$$

where $\Omega \equiv \frac{1}{2}(e^a\Gamma_a + \frac{1}{2}\omega^{ab}\Gamma_{ab} + \frac{1}{5!}b^{abcde}\Gamma_{abcde})$ is an $sp(32)$ connection. The bosonic part of (56) can be written as

$$L_{11}^{sp(32)}(\Omega) = 2^{-6}L_{G\,11}^{AdS}(\omega,e) - \frac{1}{2}L_{T\,11}^{AdS}(\omega,e) + L_{11}^b(b,\omega,e).$$

The fermionic Lagrangian is

$$\begin{aligned}L_F = &6(\bar{\psi}R^4 D\psi) - 3\left[(D\bar{\psi}D\psi) + (\bar{\psi}R\psi)\right](\bar{\psi}R^2 D\psi)\\ &-3\left[(\bar{\psi}R^3\psi) + (D\bar{\psi}R^2 D\psi)\right](\bar{\psi}D\psi) + \\ &2\left[(D\bar{\psi}D\psi)^2 + (\bar{\psi}R\psi)^2 + (\bar{\psi}R\psi)(D\bar{\psi}D\psi)\right](\bar{\psi}D\psi),\end{aligned}$$

where $R = d\Omega + \Omega^2$ is the $sp(32)$ curvature. The supersymmetry transformations (43) read

$$\delta e^a = \tfrac{1}{8}\bar{\epsilon}\Gamma^a\psi \qquad \delta\omega^{ab} = -\tfrac{1}{8}\bar{\epsilon}\Gamma^{ab}\psi$$

$$\delta\psi = D\epsilon \qquad \delta b^{abcde} = \tfrac{1}{8}\bar{\epsilon}\Gamma^{abcde}\psi.$$

Standard eleven-dimensional supergravity [41] is an N=1 supersymmetric extension of Einstein-Hilbert gravity that cannot accomodate a cosmological constant [48,49]. An $N > 1$ extension of this theory is not known. In our case, the cosmological constant is necessarily nonzero by construction and the extension simply requires including an internal $so(N)$ gauge field coupled to the fermions, and the resulting Lagrangian is an $osp(32|N)$ CS form [34].

7 Discussion

The supergravities presented here have two distinctive features: The fundamental field is always the connection **A** and, in their simplest form, they are pure CS systems (matter couplings are discussed below). As a result, these theories possess a larger gravitational sector, including propagating spin connection. Contrary to what one could expect, the geometrical interpretation is quite clear, the field structure is simple and, in contrast with the standard cases, the supersymmetry transformations close off shell without auxiliary fields.

A. Torsion. It can be observed that the torsion Lagrangians (L_T)are odd while the torsion-free terms (L_G) are even under spacetime reflections. The minimal supersymmetric extension of the AdS group in $4k-1$ dimensions requires using chiral spinors of $SO(4k)$ [20]. This in turn implies that the gravitational action has no definite parity, but requires the combination of L_T and L_G as described above. In $D = 4k+1$ this issue doesn't arise due to the vanishing of the torsion invariants, allowing constructing a supergravity theory based on L_G only, as in [13]. If one tries to exclude torsion terms in $4k-1$ dimensions, one is forced to allow both chiralities for $SO(4k)$ duplicating the field content, and the resulting theory has two copies of the same system [52].

B. Field content and extensions with N>1. The field content compares with that of the standard supergravities in $D = 5, 7, 11$ as follows:

D	Standard supergravity	CS supergravity
5	$e^a_\mu\ \psi^\alpha_\mu\ \psi_{\alpha\mu}$	$e^a_\mu\ \omega^{ab}_\mu\ \psi^\alpha_\mu\ \psi_{\alpha\mu}\ b$
7	$e^a_\mu\ A_{[3]}\ \psi^{\alpha i}_\mu\ a^i_{\mu j}\ \lambda^\alpha\ \phi$	$e^a_\mu\ \omega^{ab}_\mu\ \psi^{\alpha i}_\mu\ a^i_{\mu j}$
11	$e^a_\mu\ A_{[3]}\ \psi^\alpha_\mu$	$e^a_\mu\ \omega^{ab}_\mu\ \psi^\alpha_\mu\ b^{abcde}_\mu$

Standard supergravity in five dimensions..... The theory obtained with our scheme is the same one discussed by Chamseddine in [13].

Standard seven-dimensional supergravity is an $N = 2$ theory (its maximal extension is N=4), whose gravitational sector is given by Einstein-Hilbert gravity with cosmological constant and with a background invariant under $OSp(2|8)$ [46,47]. Standard eleven-dimensional supergravity [41] is an N=1 supersymmetric extension of Einstein-Hilbert gravity that cannot accomodate a cosmological constant [48,49]. An $N > 1$ extension of this theory is not known.

In the case presented here, the extensions to larger N are straighforward in any dimension. In $D = 7$, the index i is allowed to run from 2 to $2s$, and the Lagrangian is a CS form for $osp(2s|8)$. In $D = 11$, one must include an internal $so(N)$ field and the Lagrangian is an $osp(32|N)$ CS form [1]. The cosmological constant is necessarily nonzero in all cases.

C. Spectrum. The stability and positivity of the energy for the solutions of these theories is a highly nontrivial problem. As shown in Ref. [4], the number of degrees of freedom of bosonic CS systems for $D \geq 5$ is not constant throughout phase space and different regions can have radically different dynamical content. However, in a region where the rank of the symplectic form is maximal the theory behaves as a normal gauge system, and this condition is stable under perturbations. As it is shown in [50], there exists a nontrivial extension of the AdS superalgebra with one abelian generator for which anti-de Sitter space without matter fields is a background of maximal rank, and the gauge superalgebra is realized in the Dirac brackets. For example, for $D = 11$ and $N = 32$, the only nonvanishing anticommutator reads

$$\{Q^i_\alpha, \bar{Q}^j_\beta\} = \frac{1}{8}\delta^{ij}\left[C\Gamma^a J_a + C\Gamma^{ab} J_{ab} + C\Gamma^{abcde} Z_{abcde}\right]_{\alpha\beta}$$
$$-M^{ij} C_{\alpha\beta},$$

where M^{ij} are the generators of $SO(32)$ internal group. On this background the $D = 11$ theory has 2^{12} fermionic and $2^{12} - 1$ bosonic degrees of freedom. The (super)charges obey the same algebra with a central extension. This fact ensures a lower bound for the mass as a function of the other bosonic charges [51].

D. Classical solutions. The field equations for these theories in terms of the Lorentz components (ω, e, b, a, ψ) are spread-out expressions for $<\mathbf{F}^{n-1}G_{(a)}>= 0$, where $G_{(a)}$ are the generators of the superalgebra. It is rather easy to verify that in all these theories the anti-de Sitter space is a classical solution , and that for $\psi = b = a = 0$ there exist spherically symmetric, asymptotically AdS standard [25], as well as topological [36] black holes. In the extreme case these black holes can be shown to be BPS states.

E. Matter couplings. It is possible to introduce a minimal couplings to matter of the form $\mathbf{A}\cdot\mathbf{J}$. For $D = 11$, the matter content is that of a theory with (super-) 0, 2, and 5–branes, whose respective worldhistories couple to the spin connection and the b fields.

F. Standard SUGRA. Some sector of these theories might be related to the standard supergravities if one identifies the totally antisymmetric part of the contorsion tensor in a coordinate basis, $k_{\mu\nu\lambda}$, with the abelian 3-form, $A_{[3]}$. In 11 dimensions one could also identify the antisymmetric part of b with an abelian 6-form $A_{[6]}$, whose exterior derivative, $dA_{[6]}$, is the dual of $F_{[4]} = dA_{[3]}$. Hence, in $D = 11$ the CS theory possibly contains the standard supergravity as well as some kind of dual version of it.

Acknowledgments

The authors are grateful to R. Aros, M. Bañados, O. Chandía, M. Contreras, A. Dabholkar, S. Deser, G. Gibbons, A. Gomberoff, M. Günaydin, M. Henneaux, C. Martínez, F. Méndez, S. Mukhi, R. Olea, C. Teitelboim and E. Witten for many enlightening discussions and helpful comments. This work was supported in part by grants 1960229, 1970151, 1980788 and 3960009 from FONDECYT (Chile), and 27-953/ZI-DICYT (USACH). Institutional support to CECS from Fuerza Aérea de Chile and a group of Chilean private companies (Business Design Associates, CGE, CODELCO, COPEC, Empresas CMPC, Minera Collahuasi, Minera Escondida, NOVAGAS and XEROX-Chile) is also acknowledged.

References

1. R. Troncoso and J. Zanelli, Phys. Rev. **D 58** R101703, (1998).
2. R. Troncoso and J. Zanelli, *Gauge Supergravities for all Odd Dimensions*, lecture presented at the Third Meeting Quantum Gravity in the Southern Cone, Bariloche, January 1998. hep-th/9807029.
3. M. Henneaux, Phys. Rep. **126** (1985) 1.

4. M. Bañados, L. J. Garay and M. Henneaux, Phys. Rev.**D53** R593 (1996); Nucl. Phys. **B476** 611 (1996).
5. C. Lanczos, Ann. Math. **39** (1938) 842.
6. D. Lovelock, J. Math. Phys. **12** (1971) 498.
7. B. Zumino, Phys. Rep. **137** (1986) 109.
8. A. Mardones and J. Zanelli, Class. Quantum Grav. **8** (1991) 1545.
9. P. K. Townsend, *Three Lectures on Quantum Supersymmetry and Supergravity*, Trieste Summer School '84, B. de Wit, P. Fayet, and P. van Nieuwenhuizen, editors.
10. J. C. Taylor and V. O. Rivelles, Phys. Lett. **B104** (1981) 131; **B121** (1983) 38.
11. E. Witten, Nucl. Phys. **B311** (1988) 46.
12. A. Chamseddine, Phys. Lett. **B233** (1989) 291.
13. A. Chamseddine, Nucl.Phys. **B346** (1990) 213.
14. M. Bañados, C. Teitelboim and J. Zanelli, Phys. Rev. **D49** (1994) 975.
15. J. Zanelli, Phys. Rev. **D51** (1995) 490.
16. M. Bañados, R. Troncoso and J. Zanelli, Phys. Rev. **D54** (1996) 2605.
17. P. van Nieuwenhuizen, Phys. Rep. **68** (1981) 1.
18. J. W. van Holten and A. Van Proeyen, J. Phys. **A 15** (1982) 3763.
19. M. Sohnius, Phys. Rep. **28** (1985) 39.
20. M. Günaydin and C. Saclioglu, Comm. Math. Phys. **87** (1982) 159.
21. T. Regge, Phys. Rep. **137** (1986) 31.
22. C. Teitelboim and J. Zanelli, Class. and Quantum Grav. **4** (1987) L125.
23. B. Zwiebach, Phys. Lett. **156B** (1985) 315.
24. D. G. Boulware and S. Deser, Phys. Rev. Lett. **55** (1985) 2656.
25. M. Bañados, C. Teitelboim and J. Zanelli, *Lovelock- Born-Infeld Theory of Gravity* in *J. J. Giambiagi Festschrift*, H. Falomir, E. Gamboa-Saraví, P. Leal, and F. Schaposnik (eds.), World Scientific, Singapore, 1991.
26. J.T. Wheeler, Nucl. Phys. **B268** (1986) 737; **B273** (1986) 732. B. Whitt, Phys. Rev. **D38** (1988) 3001. R. C. Myers and J. Simon, Phys. Rev. **D38** (1988) 2434. D.L. Wiltshire, *ibid.*, **38** (1988) 2445.
27. M. Henneaux, C. Teitelboim and J. Zanelli, *Gravity in Higher Dimensions*, in SILARG V, M. Novello, (ed.), World Scientific, Singapore, 1987; Phys. Rev. **A 36** (1987) 4417.
28. R. Hojman, C. Mukku and W.A. Sayed, Phys.Rev. **D22** (1980) 1915.
29. M. Contreras and J. Zanelli, *A note on the spin connection formulation of gravity* (to appear).
30. P. Freund, *Introduction to Supersymmetry* Cambridge University Press, Cambridge, U.K., 1989.
31. R. Debever, (ed.), *Elie Cartan – Albert Einstein, Lettres sur le Par-*

allélisme Absolu, 1929-1932 Académie Royal de Belgique & Princeton University Press (1979).
32. M. Nakahara, *Geometry, Topology and Physics*, Adam Hilger, New York, 1990. T. Eguchi, P.B. Gilkey and A.J. Hanson, Phys. Rep. **66** (1980) 213.
33. S.W. MacDowell and F. Mansouri, Phys. Rev. Lett.**38** (1977) 739; Erratum-ibid.**38** (1977) 1376.
34. R. Troncoso, Doctoral Thesis, University of Chile (1996).
35. O. Chandía and J.Zanelli, *Torsional Topological Invariants (and their relevance for Real Life)*. Lecture given at La Plata Meeting on Trends in Theoretical Physics, La Plata, Argentina, May 1997, hep-th/9708138.
36. S. Aminneborg, I. Bengtsson, S. Holst and P. Peldan, Class. Quantum Grav.**13** (1996) 2707. M. Bañados, Phys. Rev. **D57** (1998), 1068. R.B. Mann, *Topological Black Holes: Outside Looking In*, gr-qc/9709039.
37. R. Aros, R. Olea, R. Troncoso and J. Zanelli, *Constant Curvature Black Branes* (manuscript in preparation).
38. R.Cai and K.Soh, *Topological Black Holes in Dimensionally Continued Gravity*, gr-qc/9808067
39. C. Martínez, R. Troncoso and J. Zanelli (manuscript in preparation).
40. F. Müller-Hoissen, Nucl.Phys. **B346** 235 (1990)
41. E. Cremmer, B. Julia and J. Scherk, Phys. Lett. **76B**(1978) 409.
42. W. Nahm, Nucl. Phys. **B135** (1978) 149. J. Strathdee, Int. J. Mod. Phys. **A 2** (1987) 273.
43. J. A. de Azcárraga, J. P. Gauntlet, J. M. Izquierdo and P. K. Townsend, Phys. Rev. Lett. **63** (1989) 2443.
44. P. K. Townsend, *p-Brane Democracy*, hep-th/9507048
45. H. Nishino and S. J. Gates, Phys. Lett. **B388** (1996) 504.
46. P.K. Townsend and P. van Nieuwenhuizen, Phys. Lett. **125B** (1983) 41.
47. A. Salam and E. Sezgin, Phys. Lett. **126B** (1983) 295.
48. K. Bautier, S. Deser, M. Henneaux and D. Seminara, Phys. Lett. **B406**,(1997) 49.
49. S. Deser, *Uniqueness of D=11 Supergravity*, Lecture presented at this meeting, August 1997, hep-th/9712064.
50. O. Chandía, R. Troncoso and J. Zanelli (in preparation).
51. G. W. Gibbons and C. M. Hull, Phys. Lett. **109B**, 190 (1982).
52. P. Horava, *M-Theory as a Holographic Field Theory*, hep-th/9712130.